OCEANX

OUR FUTURE BELOW

"The sea, once it casts its spell, holds one in its net of wonder forever."

—**JACQUES COUSTEAU**, oceanographer and storyteller

"Watching Jacques Cousteau's ocean explorations was thrilling. Because of Cousteau, I started diving and took my son Mark diving, and he also fell in love with it. I came to see that the ocean is our most important and least explored natural resource, larger than all land combined. That led Mark and me to found OceanX to explore the ocean and bring it back to people through great media."

—**RAY DALIO,** cofounder of OceanX

OCEANX
OUR FUTURE BELOW

The Game-Changing Science of Ocean Exploration

JOIN THE OCEANX MISSION TO EXPLORE THE OCEAN AND TELL ITS STORY TO THE WORLD

HUGH FRANCIS ANDERSON

Afterword by MARK DALIO and VINCENT PIERIBONE

EARTH AWARE

SAN RAFAEL • LOS ANGELES • LONDON

16
12.8t
OCEANXPLORER
H
16
16

ST-253 XR
PLORER

CONTENTS

"Humanity's future depends on a healthy ocean, which is now at great risk. We decided to help people explore, love, and protect it. Together, we can secure a successful future for the ocean—and ourselves."

—MARK DALIO

PIPA

Mark and Ray Dalio founded OceanX to explore the ocean and tell its story to the world. Inspired by Jacques Cousteau's work, the organization has set out to transform scientific and public understanding. Here are just a few of the amazing missions and media projects OceanX has accomplished so far:

2012 Captured the first-ever footage of a giant squid in its natural habitat near Japan's Ogasawara Islands.

2014 First to discover the biofluorescence in more than 180 species of fish.

2016 Explored the ocean's blue holes for the Emmy Award–winning television series *Years of Living Dangerously*.

2018 Completed the deepest manned-submersible dive ever in Antarctica, to 1,000 in the South Shetland Trench, for the internationally acclaimed, landmark nature documentary series *Blue Planet II* with BBC Studios.

2019 Tagged a bluntnose sixgill shark from a submersible for the first time in scientific history, a feat completed at more than 500 meters deep that has been viewed more than 27 million times on YouTube.

2023 Released *OceanXplorers*, an epic six-part underwater discovery series filmed aboard the *OceanXplorer*, on National Geographic, Disney+, and Hulu.

2024 Produced the first ever livestream from 350 meters below the ocean's surface by connecting marine biologist Diva Amon from a submersible with a live audience in Davos.

2025 Reached more than 2.2 billion total social media impressions across Facebook, Instagram, LinkedIn, TikTok, X, and YouTube, engaging a global audience of ocean lovers.

INTRODUCTION

"Ocean exploration reveals the ocean's mysteries. Scientific research builds humanity's understanding of its potential. OceanX and the *OceanXplorer* exist to bring these wonders to light for the world in a way that fuels an inspiring, bright future that we can only imagine today."

—VINCENT PIERIBONE

The ocean is our lifeblood. It covers almost 71 percent of the Earth's surface, contains 97 percent of its water, and is home to most of the planet's biodiversity. Yet, despite its enormous size, we have explored just 5 percent of it, and it continues to face threats from overconsumption, pollution, and neglect. But the tides finally appear to be changing. "We're beginning to grasp the significance that the most important thing we take from the ocean is our existence," says renowned marine biologist and explorer Dr. Sylvia Earle. "Life as we know it is dependent on maintaining the integrity of the ocean. No ocean, no life. The ocean makes life habitable."

Our need for and connection to the ocean goes back to the earliest years of humanity. Homo sapiens began traveling by sea some 60,000 years ago, with the migration to Australia from Southeast Asia. The earliest known artifacts demonstrating our reliance on the ocean for sustenance were found in the depths of Sakitari Cave on Okinawa Island, Japan. Among the skeletons, beads, and scrapers, archaeologists unearthed fishing hooks delicately carved from the shells of sea snails. Using radiocarbon dating methods on charcoal found in the same layer, they estimated the hooks to be almost 23,000 years old, making them the oldest known fishing technology. This deep history reveals that our relationship with the ocean is in the very fabric of what it means to be human.

As technologies advanced, so too did our curiosity for what lay beyond the horizon and beneath the surface. Shipbuilding allowed us to reach distant lands and, in doing so, to discover species we could harvest not only for nourishment but also for economic gain. By the sixteenth and seventeenth centuries, written accounts of fish and whale stocks offered extensive documentation of the ocean's biodiversity, even as we plundered many species to the brink of extinction. The inventions of the diving bell and the first submarine in the seventeenth century marked the birth of underwater exploration. The evolution of undersea technology was rapid. By the early nineteenth century, submarines were an important military technology, and the copper diving helmet heralded the beginning of scuba diving. With these technologies, humans could explore an entirely new environment—the underseas world—for the very first time.

Pages 10–11: OceanX's science submersible, *Neptune*, prepares for launch. Diver swims with manta rays in the Azores.

Pages 12–13: *OceanXplorer*, the most advanced research vessel ever built, exploring the fjords of Norway.

Right: Coral in Micronesia.

Pages 16–17: An OceanX submersible on a dive in Antarctica waters.

"The sea, the great unifier, is man's only hope. Now, as never before, the old phrase has a literal meaning: We are all in the same boat."

—**JACQUES COUSTEAU**, oceanographer and storyteller

Throughout the Age of Exploration, there was extensive mapping of the ocean surface and surrounding coastlines, but scientific study was still confined to land. The Italian scientist Luigi Ferdinando Marsili is widely considered the father of modern oceanography because of his eighteenth-century investigation of the biology of the Gulf of Lion, but it was not until the late nineteenth century that dedicated oceanographic missions began. Inspired by Charles Darwin's work on his 1831–1836 circumnavigation of the globe, the British launched HMS *Challenger* on the first scientific expedition of the physical, chemical, geological, and biological conditions of the ocean in 1872. Covering almost 70,000 nautical miles over four years, the expedition documented and catalogued thousands of new species and observations. "The Challenger Expedition has played a very large part in all the recent advances in oceanographical knowledge," wrote the naturalist Sir John Murray in the expedition report published in 1895. "[This expedition] mark[s] the greatest advance in the knowledge of our planet since the celebrated geographical discoveries of the fifteenth and sixteenth centuries."

Right: A manta ray spotted in the Azores.

Conflict is one of the greatest drivers for technological advancement, and the Second World War brought tremendous innovations in ocean technologies that leaped from the military sphere into the scientific. Sonar, for example, which was designed to detect submarines, also enabled scientists to map ocean floors and basins and revolutionized the accuracy and speed with which they could collect data. Diving technology also boomed with the creation of the Aqua-Lung by Émile Gagnan and Jacques Cousteau, the latter of whom would personally change the face of ocean exploration and inspire a generation of conservationists.

Cousteau began making films in the 1940s about the underexplored deep ocean. From then until his passing in 1997, he produced a vast library of films and books exploring the ocean, always with an emphasis on the importance of conservation. He was awarded three Oscars, an Emmy, and the Palme d'Or for his film projects, many of which took place on his ship, *Calypso*. No single person has made a greater cultural impact on our desire to document and preserve the ocean than Cousteau. He showed the world how great media, in combination with groundbreaking scientific research, can drive global awareness and, ultimately, change.

Not long after Cousteau began captivating audiences with the undersea world, the success of the 1969 Apollo 11 Moon mission brought the wonder of space to a global audience. In the years since, we have sent probes or manned missions to five extraterrestrial planets and three moons. The wonder that space travel invokes, and the sense of hope it instills, has inspired a wave of media, marketing, and storytelling. And so, over time, the public imagination shifted away from Earth and into space, and along with it, significant funding.

Right: Sperm whales in the Azores.

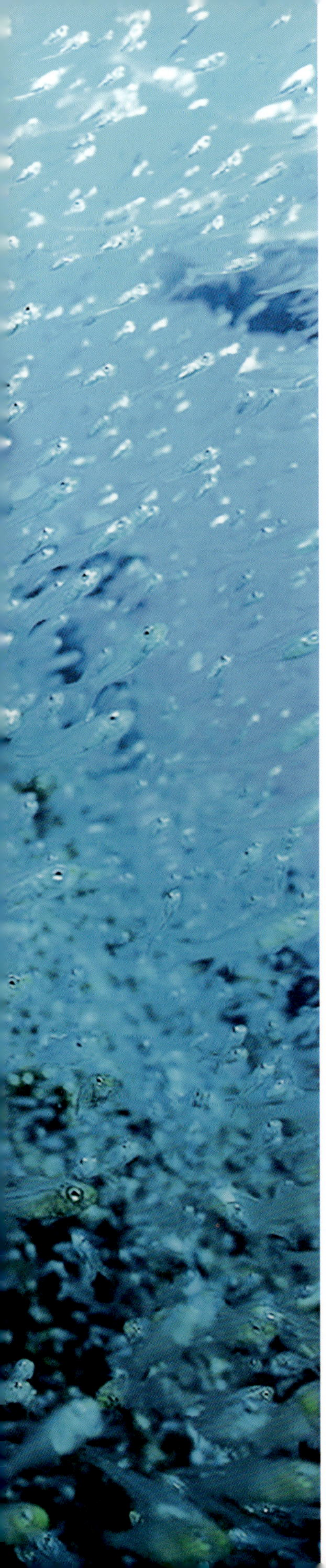

Left: Thriving reef ecosystems at Aldabra Atoll, Seychelles.

This has been especially true in the United States, where organizations like the National Aeronautics and Space Administration (NASA) have captivated the public's imagination, whereas the National Oceanic and Atmospheric Administration (NOAA) remains lesser known to the public. Looking globally, and including non-government spending, the ratio between space exploration spending and ocean spending is roughly 120 to 1. This is a shocking statistic given the importance of ocean understanding and exploration for the health of the planet and humanity's continued existence on it.

More people get their meals from the ocean than from anywhere else, shipping by sea accounts for the transport of 80 percent of the world's goods, and ocean-related industries employ millions of people. New breakthrough medicines, including anticancer drugs and painkillers, have rekindled the pharmacology industry's interest in microbes from the ocean. Marine invertebrates that dwell in the deep sea, such as sea sponges and sea snails, have been shown to produce compounds that can effectively fight a whole host of illnesses. The vulnerability of these soft-bodied creatures requires them to produce compounds that can fight off bacteria and predators alike. Ziconotide, for example, is an analgesic discovered to be an effective, nonaddictive treatment for severe and chronic pain. It is derived from a neurotoxin peptide used by the cone snail to paralyze its prey. Other marine-derived medicines have proven effective at treating everything from Alzheimer's to cancer. The effort to develop these drugs has increased drastically in recent years, and the marine pharmaceutical industry is expected to see an almost 10 percent growth year on year for the next decade.

As new illnesses arise, the need for new discoveries intensifies. The COVID-19 pandemic provides a relevant recent example, since the most frequently used test to detect COVID-19, the PCR test, was derived from a water-based discovery: a microbe found in the harsh environment of a hot spring in Yellowstone National Park in the 1960s.

The ocean, particularly the deep sea, presents numerous inhospitable environments where the forces of natural selection have shaped one-of-a-kind adaptations that have the potential to create new treatments and new technologies. For example, the deep ocean is replete with creatures that use bioluminescence—a chemical reaction to produce light—to communicate in the dark. It occurs naturally in living things such as fireflies, anglerfish, and plankton, and scientists have used bioluminescent proteins to help solve a daunting problem in neuroscience: studying the human brain without harming the person to whom it belongs.

DAMEN
MORELLO
MORELLO
MORELLO

"We need to respect the ocean and take care of [it] as if our lives depended on it. Because they do."

—DR. SYLVIA EARLE, marine biologist and explorer

The ocean also plays an essential role in regulating Earth's climate. More than 90 percent of excess anthropogenic heat is absorbed by the ocean. Diatoms, seagrass, and other marine organisms absorb more carbon than terrestrial systems. And there's an enormous trove of unused energy in the ocean that could be harnessed with little negative impact on the environment, which could help reduce our reliance on fossil fuels. Wind turbines are significantly more productive when positioned in the ocean because there are no landmasses to block the wind. Thermal gradient energy uses the difference in temperature between the warmer surface and the much colder depths to spin turbines. And wave energy, which utilizes the natural motion of waves and tides, and green hydrogen, which creates hydrogen power by splitting seawater atoms, both show potential as technologies advance.

Taking all of this into account, the question is not whether the ocean—and the exploration and discovery that helps us realize the full potential of our relationship with it—deserves a tremendous amount of attention and public support. The question is how to generate it. The Apollo mission supercharged space exploration decades ago, but until recently the ocean was still waiting for its own moon shot. So, in 2018, Ray and Mark Dalio created OceanX to do just that.

OceanX is a mission to support scientists to explore the ocean and to bring it back to the world through captivating media. Uniting leading media, science, and philanthropy partners, OceanX utilizes next-generation technology, fearless science, compelling storytelling, and immersive experiences to educate, inspire, and connect the world with the ocean and build a global community deeply engaged with understanding, enjoying, and protecting the ocean.

"The purpose of OceanX is to support exploration and show it to the world," says Ray Dalio, founder of and advisor to OceanX. "We expect our efforts to reach somewhere between half a billion to a billion people, and we hope that they will be inspired by what they see so that they learn to love and treat our ocean well."

This commitment to uncovering scientific knowledge and inspiring the world with its potential runs through OceanX's every activity as it works to drive positive change for the ocean. *OceanXplorer*, a technological marvel when it comes to deep-sea exploration, is the platform that brings this vision to life, supporting scientific research, educational programming, and immersive storytelling, all of which make data about the ocean more accessible while also showcasing its innate magic and wonder.

Pages 24–25: OceanX's media submersible, *Nadir*, prepares to exit the sub hangar and to launch.

Left: Flourishing coral reef in the Red Sea.

Page 28: A vibrant, healthy coral reef provides shelter and food for several different marine species in the Red Sea.

Page 29: A pod of dolphins spotted in the Red Sea.

IS TAPPING INTO OUR INNATE LOVE OF THE OCEAN A KEY TO OUR WELLBEING?

There is no doubt that our physical existence on Earth depends on the ocean. But are there also spiritual and emotional aspects of our connection to it? The esteemed naturalist and evolutionary biologist E. O. Wilson introduced the concept of "biophilia" in his 1984 book of the same name. It theorizes that humans have an ingrained love of nature that is imprinted on our DNA due to our long reliance on and need for nature throughout our evolution. "From infancy we concentrate happily on ourselves and other organisms. We learn to distinguish life from the inanimate and move toward it like moths to a porch light," writes Wilson in *Biophilia*. "To explore and affiliate with life is a deep and complicated process in mental development. To an extent still undervalued in philosophy and religion, our existence depends on this propensity, our spirit is woven from it, hope rises on its currents. . . . To the degree that we come to understand other organisms, we will place a greater value on them, and on ourselves."

Other researchers have expanded on Wilson's theory, looking to explain the various factors at play in nature's positive impacts on human mental health and well-being. In a 2013 paper titled "Happiness Is Greater in Natural Environments," George Mackerron and Susana Mourato outline three commonly cited reasons for linking natural environments to positive health outcomes. First, consistent with the biophilia hypothesis, there appear to be direct neurological pathways honed by evolution that bring stress

Right: Diver holds up a sea star in Alaska.

reduction when in natural places. Second, nature has fewer stressors, such as noise and air pollution, than other facets of our lives, which results in a greater sense of health. Third, natural environments promote activity, even if only as basic as physical movement, which is directly linked to positive mental health. These factors help establish why we are not only drawn to natural spaces but why we thrive in them, too. As Mackerron and Mourato write, "Amongst study participants, happiness is greater in natural environments, even after controlling for a wide range of potential confounders."

Dr. Matthew White, a specialist in environmental psychology and a senior lecturer at the University of Exeter, reviewed the scientific literature on the impact of aquatic environments ("blue spaces") on many elements of human health, including mental well-being. In a paper White coauthored titled "Blue Space, Health and Well-being: A Narrative Overview and Synthesis of Potential Benefits," he describes that while being around water has many of the same impacts as being in nature, it may also have more subtle and unique influences, especially on reducing anxiety and increasing feelings of relaxation—a notion that will feel true and commonsensical to anyone who has ever sat at the ocean's edge, listening to the waves and staring out to the blue horizon.

Above: School of fish off the coast of Fernando de Noronha in Brazil. Sea of Cortez.

Right: An enormous school of fish swim by in Raja Ampat, Indonesia.

Left: Whale shark opens its mouth wide in the Sea of Cortez.

PART I:

THE DAWN OF THE *OCEANXPLORER*

SWL-9.5TONS AT 3.74M
ALUCIA
ALUCIA
KINGSTOWN
IMO #73-7823

ALUCIA
MAJURO

ALUCIA: THE JOURNEY TO OCEANX

Growing up along Long Island's beaches in the 1950s and 1960s, a young Ray Dalio was captivated by Jacques Cousteau's undersea adventures, which brought the ocean environment to the broader public for the first time. Cousteau's documentaries and TV programs instilled in Ray a passion for ocean exploration. They inspired him to become a scuba diver and teach his sons diving and years later—after building his investment business, Bridgewater Associates, into the world's largest hedge fund—to commit to a new mission of doing for others what Cousteau did for him through the OceanX initiative.

The story of how OceanX came to be flows naturally from Ray's passion for the ocean and his desire to share it with his son, Mark. Mark had learned to dive with Ray as a teenager, and after college he found his way to National Geographic, where he worked in nature documentary filmmaking, which further developed his interest in the ocean. Around the same time, Ray and his family were expanding their philanthropic activities, beginning with the founding of Dalio Philanthropies in 2003. In 2011, they signed the Giving Pledge and committed to give away the majority of their wealth to charitable causes.

Pages 36–37: Alucia launches a submersible in New Zealand.

Left: Alucia in the icy waters surrounding Antarctica.

> “The intersection between ocean health and humanity’s health is crucial for our survival.”
>
> —**RAY DALIO**, cofounder of OceanX

Given his lifelong love of the ocean, marine science and conservation were key parts of Ray’s giving from the start, and especially through his support of the Woods Hole Oceanographic Institution (WHOI), one of the world’s leading marine institutes. In 2012, Ray purchased and refit a 56-meter (184-foot) former French marine research and reconnaissance vessel named *Alucia*. In partnership with WHOI, he equipped it with the latest in marine science technology, including a launch and recovery platform for diving and submersible operations. Working closely with Mark, he set out to enhance the ship’s media capabilities.

Mark’s professional passion lay in filmmaking and storytelling. Working at the National Geographic Society to document wildlife across the globe, he had observed two things: Natural history productions often just focus on individual projects before moving on to something else; and scientists struggle with communicating their work. “I’ve always loved the oceans, and I saw an opportunity to mesh the two disciplines, where we’d be able to do ocean exploration and interesting science, but then also have the

Right: An OceanX submersible captures biodiversity in Raja Ampat, Indonesia.

ability to document and tell that story in a range of ways," says Mark. "We were equipped right from the beginning to tell the stories of the scientific work that we were doing and to extend that knowledge to further future projects and to create education content."

Through WHOI and the purchase of *Alucia*, Ray also met filmmaker and fellow ocean explorer James Cameron, who had recently completed his *Deepsea Challenge* expedition. Piloting a custom-built submersible, Cameron descended to the bottom of Challenger Deep in the southern end of the Mariana Trench in the Pacific Ocean. At 10,935 meters (35,876 feet), it is the deepest place on Earth. "I've always dreamed of diving to the deepest place in the ocean. For me it went from a boyhood fantasy to a real quest, like climbing Everest, as I learned more about deep-ocean exploration and became an explorer myself in real life," says Cameron. "This quest was not driven by the need to set records, but by the same force that drives all science and exploration—curiosity."

United in their pursuits and with complementary skill sets, Cameron joined forces with the Dalios to advise on how to use the ship and its media abilities most effectively. Their shared mission was simple: to conduct daring science and present it to the world in the most stunning and captivating way possible. "There's an intellectual part to exploring the ocean, and there's an emotional part. Intellectually, the ocean is our biggest, most important asset. It's bigger than all the continents combined and contains 90 percent of habitable space, but it's almost completely unexplored. So it's logical to support exploration," says Ray Dalio. "And emotionally, it's fabulous. When you go down there and see all the amazing species and environments, it's beautiful and exciting and you just want to share that with the world."

Left: Large coral photographed by divers in Raja Ampat, Indonesia.

ALUCIA: THE PROOF OF A CONCEPT

Summer 2012, Japan's Ogasawara Islands

As the saying goes, timing is everything, and the launch of *Alucia* coincided perfectly with a film project Mark was developing in partnership with the Discovery Channel and the Japan Broadcasting Corporation (NHK): a mission to help a team of scientists document the elusive giant squid.

For centuries, seafarers around the globe had spread the myth of the kraken, a squidlike monster capable of devouring entire ships. And though these depictions were highly exaggerated, enough tentacle fragments had turned up in fishermen's bycatch to make clear that the giant squid was, in fact, lurking down below. For Dr. Tsunemi Kubodera, who in 2004 was the first person to photograph a giant squid in its natural habitat, there was little doubt that the creature could grow up to 12 meters (40 feet), with tentacles longer than the height of four average-sized people. But learning more about the elusive species was daunting, since they dwell 1,000 meters (3,300 feet) below the surface. It was a perfect challenge for *Alucia* and its deep-diving submersibles.

So off the ship went for the deep waters of the Pacific, with Kubodera joining the mission along with Dr. Steve O'Shea, a New Zealand–based marine biologist and giant squid expert, and Dr. Edith Widder, a renowned oceanographer and marine biologist who had developed a unique optical lure for attracting the squid, called the electronic jellyfish (e-jelly).

Pages 44–45: School of baracudas photographed near the Solomon Islands and Papua New Guinea.

Right: OceanX submersible prepares to dive on mission in Antartica.

ALUCIA

The team traveled to the Ogasawara Islands, a biodiverse archipelago to the south of Japan, where they spent more than 400 hours in the Triton submersible, using specialized ultra-low-light cameras and Dr. Widder's e-jelly. This device mimicked the defense bioluminescence of the Atolla jellyfish, successfully tricking an enormous giant squid into coming close enough to capture the world's first video footage of the creature in its natural habitat. "Had this animal had its feeding tentacles intact and fully extended, it would have been as tall as a two-story house. How could something that big live in our ocean yet remain unfilmed until now?" asks Dr. Widder. "There are great discoveries yet to be made down there—fantastic creatures representing millions of years of evolution and possibly bioactive compounds that could benefit us in ways that we can't even yet imagine."

This was the culmination of the life's work of Kubodera, O'Shea, and Widder, and it showed how imperative a state-of-the-art research and storytelling vessel could be to not only aid scientists but help show their discoveries to the world. The footage resulted in a groundbreaking documentary for both NHK and the Discovery Channel, with millions of people around the world captivated by their discovery. Ray and Mark had inspired another Cousteau-like global media moment—exactly the proof of concept they'd been looking for. The success of the giant squid mission set Alucia and the Dalios off on a worldwide tour, providing a science and media production platform for many of the most important ocean documentaries of the 2010s.

This spread: A seal lays on the ice in front of Alucia in Antarctica.

Pages 50–51: Submersible in front of Alucia prepares to dive below in the waters of Antarctica.

N988MD

Spring 2016, Gulf of Mexico

In the years that followed, the *Alucia* team helped scientists and storytellers complete nearly twenty missions. These included discoveries in the Galapagos Islands with biologist Liz Bonin, dives along the Great Barrier Reef with Sir David Attenborough, manta ray conservation in Raja Ampat, Indonesia, and groundbreaking sperm whale research in Kaikoura, New Zealand. They explored the oceans' blue holes for the Emmy-Award winning *Years of Living Dangerously*, took actor Leonardo DiCaprio underwater for his documentary *Before the Flood*, and co-produced *Oceans: Our Blue Planet* with BBC Earth, which was narrated by actor Kate Winslet.

But certainly the most notable project—both for its global media reach and its impact on ocean conservation—was working with the BBC Studios Natural History Unit on *Blue Planet II*. *Alucia* had proven to be essential in filming content featuring the Galapagos Islands and Antarctica, but it was in the Gulf of Mexico where the team captured some of the series' most iconic scenes while documenting eerie, rarely explored brine pools.

A brine pool is a highly saline underwater lake formed when ancient salt deposits seep into the ocean. The salinity, in combination with high concentrations of hydrogen sulfide and methane, make brine pools deadly for most creatures, yet they hold an abundance of biological secrets. "In the Gulf of Mexico, there are hundreds of brine pool ecosystems. We've explored less than ten," says Dr. Samantha "Mandy" Joye, professor of marine sciences at the University of Georgia and one of *Alucia*'s onboard scientists. "Every single one is unique, and that means the microbiology is unique and could have very real potential for biotechnology and medical use."

Left: Vibrant pink coral photographed in Raja Ampat, Indonesia.

Diving more than 500 meters (1,640 feet) in the *Triton* submersibles, the team was not only able to collect samples from the brine pool for further examination, but also document this primordial undersea environment with haunting imagery. "We're diving into the history of life on Earth," says Dr. Sylvia Earle, who was onboard at the time. "We have the technology to access the deep sea and know about the life that is here."

Whereas most creatures die when entering the brine pools, the team documented the hagfish, a prehistoric eellike fish, entering the brine pool for short periods to feed on carcasses. Its ability to withstand the toxicity of the pool may lead to discoveries that aid medicine, and microorganisms from similar pools have already been highlighted for potential use in anticancer drugs. "The bacteria that live in these brine pools may hold cures to diseases," notes Dr. Earle. "It's really important that we study them and understand them."

Above: The Red Sea is full of marine biodiversity, including this flourishing coral reef.

Right: OceanX encounters jellyfish on a mission in Raja Ampat, Indonesia.

Blue Planet II's global impact was immense. It was viewed by hundreds of millions of people around the world and led both to a rise in ocean conservation awareness and to the passage of ocean-protecting legislation, which was dubbed the Blue Planet Effect. With the success of these efforts, the team understood the impact and importance of their mission. But they were already beginning to see the limits of operating from a compact, fifty-year-old ship. This was partly about the physical limitations of the space—there was only so much equipment that could be crammed onboard. But there were also limitations related to retrofitting an existing ship rather than envisioning and building a new, custom vessel. Ray and Mark were also struck by the need to do something bigger and more audacious to garner even more public awareness and action. Ray often liked to compare the funding for ocean exploration to that of space exploration, which receives 100 times more funding. "The ocean is critical to the health of life on Earth, so I believe that ocean exploration is more exciting and important than space exploration," says Ray. "If we can take people into the many exciting undersea worlds, we will help to shift the funding imbalance to support what is most important here on Earth."

A new dream was born: to build upon the success of the *Alucia*, this time using a purpose-built ship, *OceanXplorer*, which would be like a Starship *Enterprise* for the oceans. "Our team of scientists, researchers, and filmmakers will explore the oceans and create exciting content and stories to inspire a human connection to the sea," says Mark. "The more that we can redirect the focus and love for space to ocean exploration and the oceans in general, the more we'll better care for and be able to discover revelations that will impact us for years to come."

Left: Submersibles prepare to dive.

Pages 58–59: Sperm whales in the Azores.

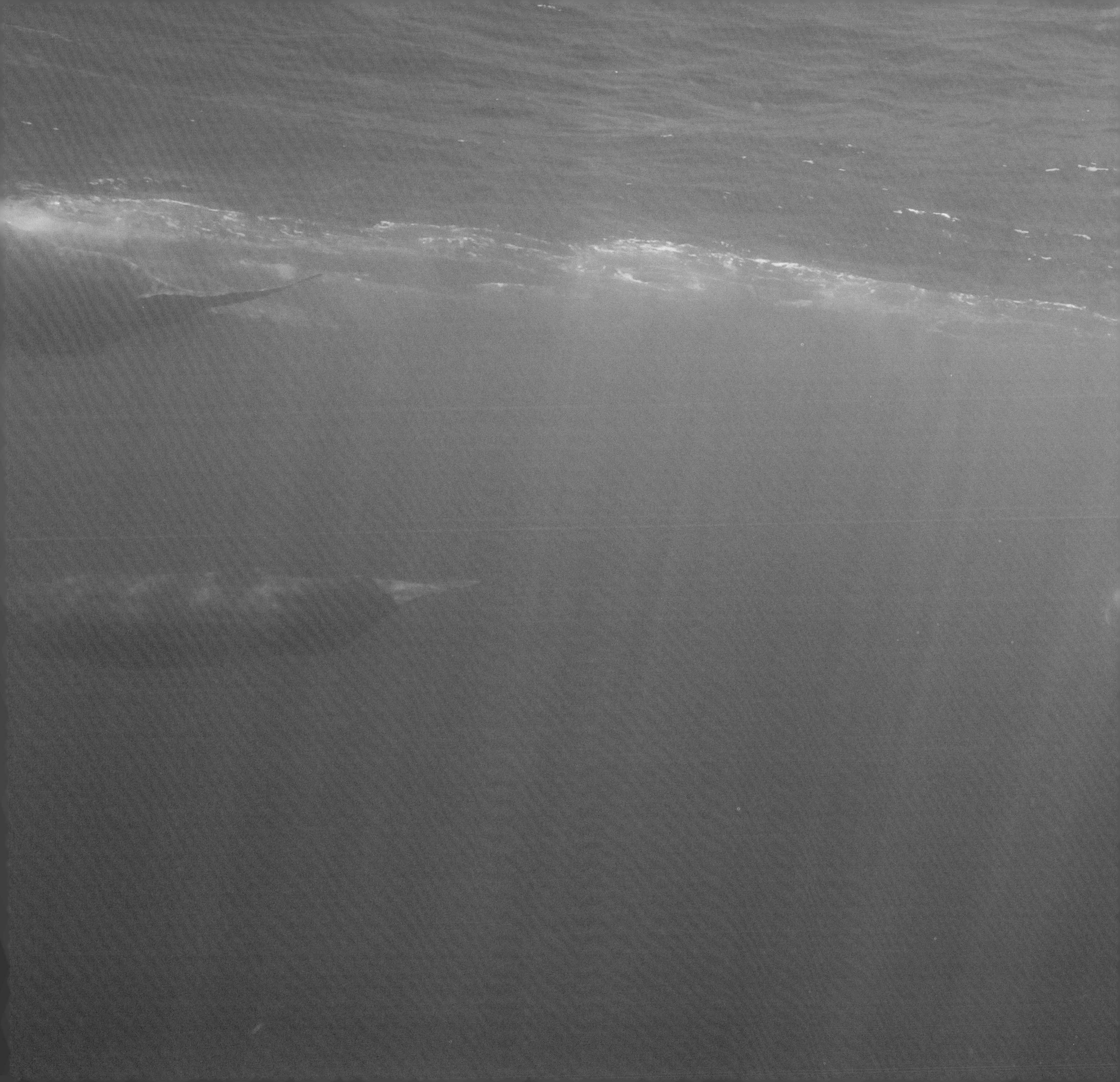

OCEANXPLORER: PIONEERING OCEAN RESEARCH AND CONNECTIVITY

"OceanX benefits from having the capacity to enlist people with talents of various sorts, whether it's filmmaking or communication or technology, to gain access to the sea, and then having a great ambassador vessel and people who can individually and collectively be a voice for the ocean. Right now it's pivotal, and why it's so important that OceanX exists."

—DR. SYLVIA EARLE, marine biologist and explorer

Right: A coral reef in Raja Ampat, Indonesia.

As a child, Vincent Pieribone was consumed by the oceans. He spent his time exploring the coastline of his home state of Florida, amazed by the wonders that lay below. "My entire life I wanted to be a marine biologist," he says. "From the moment I stepped in the water, there was never any doubt." However, when his father was diagnosed with multiple sclerosis, his focus shifted toward biology and neuroscience. "That made me rethink my approach to life, and I became a neuroscientist as a result of his illness." Now a faculty member at Lamont-Doherty Earth Observatory at Columbia University, Pieribone focuses much of his research on the medical

> "We need to do science in the ocean, we need to do adventure, and we need to do exploration, but we need to capture it all and show it to the world. While writing scientific papers is important, if we don't show the world, it's not going to have the impact that we need. If we excite the world about the oceans, we're more likely to protect it."
>
> **—VINCENT PIERIBONE,** co-CEO and chief science officer, OceanX

use of biofluorescence in understanding diseases such as AIDS, along with a host of neurological conditions. "I came across green fluorescent protein, which had just been discovered and which was groundbreaking and still is groundbreaking in its ability to visualize the brain and to visualize all of biology," he says. "The majority of green fluorescent proteins come from marine organisms—corals, jellyfish, and anemones. They come from the ocean." In 2005, Pieribone and his research partner, David F. Gruber, published *Aglow in the Dark: The Revolutionary Science of Biofluorescence*, and their research eventually led them on an expedition aboard *Alucia* to the Solomon Islands to study fluorescence in fish.

It was on this mission that Pieribone and Ray Dalio kindled a close friendship, and between 2013 and 2018, Pieribone joined Ray and *Alucia* on several expeditions, both as a researcher and an advisor. Impressed by the ambition the Dalios had for marine exploration and media, he also shared the belief that scientific research alone was not enough to ignite the public's passion for the ocean and commitment to protecting it. It needed to be presented through compelling and informative media that first and foremost inspired wonder—just as Cousteau's missions had. ""Most great science in terms of answering questions is uninteresting and longitudinal, but it's very significant, important science. And then there's media, which is a huge world of reality TV, etc., that's successful and gets to a market," says Pieribone. "What's needed is a middle ground, to do things that are upbeat—talk about how beautiful the oceans are, what's left to be discovered—and encourage young people to get into it."

Pieribone's position as a PhD scientist who shared the same mission and sensibility as the Dalios on science media made him a unique partner to help bridge the gap the Dalios were hoping to span with OceanX. Pieribone came onboard in 2016 as OceanX's new director.

Right: Clownfish in Raja Ampat, Indonesia.

The first step was securing a new ship. Ray saw an opportunity at a time when oil prices had crashed and oil exploration vessels were selling for a fraction of what they had just a few years before. These highly capable, mammoth ships were designed to explore the ocean seeking natural resources to extract, but Ray had something different in mind: converting one of them into a marine science and media platform. Doing so would be much more cost-effective and faster to complete than building from scratch, so Ray settled on the *Volstad Surveyor*, a survey ship that had been built in 2010. It would be the foundation for creating *OceanXplorer*. "We had *Alucia* and operated it successfully for several years, but we wanted something more capable," says Peter Fielding, cocaptain of *OceanXplorer*. "We've taken a ship that went from exploiting the world to exploring it for good."

Left: A Goliath grouper surrounded by a school of fish in Boynton Beach, Florida. Photo by Christian Vizl / Tandem.

PALFINGER
IMO: 9533373
CTD
ROV

With almost four times the volume of *Alucia*, *OceanXplorer* has been redesigned in consultation with leading marine scientists, the *Alucia* marine and submarine operations crews, and James Cameron, who looked over the blueprints to consider room-by-room how best to configure the ship for media capture. Page Buckner, a Hollywood set designer recommended by Cameron who had worked on the *Transformers* films, designed the on-ship graphics and wayfinding. Whereas *Alucia* only allowed for the operation of one ancillary vehicle at a time, *OceanXplorer* can simultaneously launch submarines, helicopters, an ROV, and other exploration tools. "There's so much we can do at the same time," says Roger Solem, cocaptain of *OceanXplorer*. "We can have two subs and the ROV in the water, the helicopter and drone ops in the air. It's huge and so impressive." Importantly, the ship also has berths for up to seventy people, which means more researchers and storytellers can work simultaneously. With partnerships in mind, a VIP deck was also incorporated to host partners and delegates to discuss policy making and future exploration, and to build lasting working relationships.

"At OceanX, we've long believed in and worked to harness the transformative power of media every day," says Mark Dalio, cofounder and co-CEO of OceanX. "*OceanXplorer* allows us to pair science and media together like never before and share the excitement and wonder of ocean exploration with a global audience in real time."

Left: *OceanXplorer* undergoes reparations in a Damen shipyard.

Above: The *Volstad Surveyor* in dock.

The OceanXplorer was designed to push the boundaries of ocean science and storytelling like no other vessel on the planet. Because of its unique capabilities, the OceanXplorer has achieved many firsts and discoveries, some of which are:

2020 Discovered brine pools in the Red Sea.

2020 First to observe Boulenger's anthias in the Red Sea, significantly expanding its known distribution.

2020 Discovered two new groups of shrimp and two new shrimp species in the Red Sea.

2020 Discovered a new species of black coral in the Red Sea.

2021 Discovered a new species of sea snail in the Red Sea.

2023 First to observe mesophotic corals in the Northwest Indian Ocean.

2023 First to observe and record the bramble shark in the UAE.

2024 Deepest-ever unmanned sub dive, at 4,600 meters, and manned sub dive, at 1,000 meters, in the Seychelles.

Each discovery revealed a previously unknown secret of the ocean, which was shared with the public through original media, conveying the need to protect the ocean and providing a deeper understanding of the world below the surface.

Right: *OceanXplorer* at sea in Indonesia.

PART II:

EXPLORING UNCHARTED WATERS

HELO

ST-253XR
OCEANXPLO

INSIDE *OCEANXPLORER*

"*OceanXplorer* will be the most cutting-edge, futuristic, inspirational ocean exploration ship on the water."

—RAY DALIO, cofounder of OceanX

Pages 70–71: The *OceanXplorer* on mission in Indonesia.

M/Y *OCEANXPLORER*

- LENGTH: 87.1 meters (285.8 feet)
- BEAM: 18 meters (59 feet)
- DRAFT: 6.8 meters (22.3 feet)
- DISPLACEMENT: 5,740 tons
- CRUISING SPEED: 6.2 meters per second (12 knots)
- ACCOMMODATION CAPACITY: 72 people

Above (left to right): Captains control *OceanXplorer* from the Bridge; The kitchen aboard *OceanXplorer*; Fuel for *OceanXplorer's* onboard helicopter.

Right (left to right): Navigation and communication panels in the Bridge; The crew's dining area; ROV docking deck.

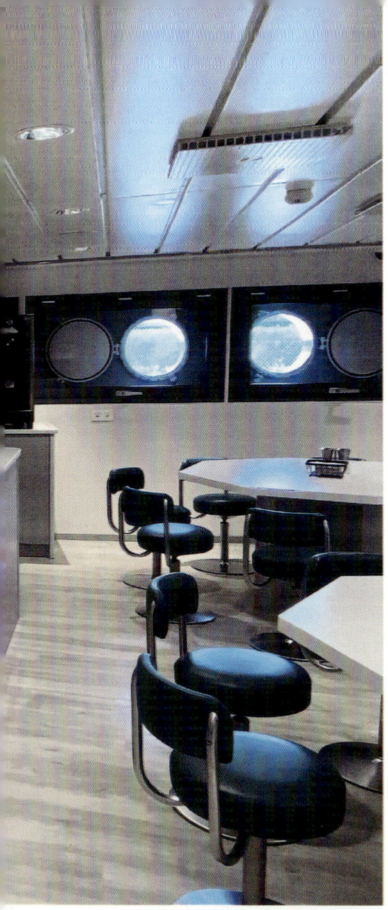

HELIFUEL

evotec
MARINE HANDLING TECHNOLOGY
OCEANX
ARGUS
XL

A JOURNEY ONBOARD

The sky is radiant. Cirrus clouds drift with the westerly breeze coming in off the North Sea. Below, a lush summer coat of pine and birch adorns the mountainous walls that rise high above Masfjorden, a fjord on Norway's western coast. As our helicopter makes its way up the otherwise empty fjord, a vessel comes into view. Ben Neel, *OceanXplorer*'s helicopter pilot, turns to me and smiles. "There's *OceanXplorer*," he says with pride. We circle in low as the warm morning light reflects off the calm water, sparkling on the gray hull. On the aft deck, a flurry of movement ensues. One of the two submersibles emerges from the hangar. The crane moves into position. A door on the starboard hull rises, the ROV appearing inside it. And as we come in to land, I see the ship's name painted across the helideck. I crane my neck at the sheer size of the vessel. It inspires a sense of wonder within me.

Right: *OceanXplorer* off the coast of Indonesia.

HELO
N988MD
16

Once off the helicopter, I step inside and descend through a warren of passageways to the aft deck, where the OceanX team is in full operation on a collaboration with a team from the University of Bergen. They are exploring the ecosystems in the depths of the fjord, expanding our knowledge of bioluminescent creatures, bluefin tuna, gelatinous plankton, mesopelagic fish, benthic organisms, and kelp forests. The Triton submersible *Nadir* is lifted into the water, followed shortly by *Neptune*. Through the 16.51-centimeter (6.5-inch)-thick acrylic domes, I observe the enchanted faces of the scientists, storytellers, and pilots within. They are about to descend to almost 500 meters (1,640 feet) for the first time, and they'll remain underwater conducting research and filming for the next six hours. This is exploration in its truest form.

I stride across the deck and pry open the heavy lever of the door labeled Sub Hangar, as if I were a steward on some intergalactic ship. Stooping through the doorway, I stare in awe at the beating heart of the vessel. The tacky smell of seawater is fused with the scents of machinery. An LED ribbon runs the length of the hangar's starboard wall, and the words *Welcome to OceanXplorer* drift by in an endless carousel. I'm struck by just how futuristic the vessel is, as if it were plucked from a science fiction movie. The creative influence of James Cameron and Buckner is plain to see.

Right: Kelp forests in Norway are an important part of the marine environment and contribute to the ocean's amazing power to hold carbon.

Each department adjoins the sub hangar. To starboard is the wet lab, where samples from the deep are brought up for initial examination. To port is the dive center, which allows scientists and storytellers to rapidly collect samples and document marine life. Just past the wet lab, farther starboard, is the CTD hanger. A turbine-like contraption used to analyze the conductivity, temperature, and depth of the water column, the CTD (conductivity, temperature, and depth) instrument cluster can be lowered to depths of almost 7,000 meters (22,966 feet). The enormous Argus ROV stands in the hanger next to it. It appears more creature than robotic vehicle, its eyes and ears giving the crew exclusive access to a world so far below the surface that we would otherwise never be able to comprehend it. I reach out and touch one of the titanium pincers. Strong enough to crush my hand, it's also capable of such delicacy that it can retrieve the most fragile samples without harm.

As I continue forward, the sound of voices grows, and I rise through an oval doorway into mission control. A host of scientists, crew, and camera operators buzzes within. Mission control is the brain of all operations, a central nervous system that controls, monitors, and documents all that happens below the surface. It is dark, illuminated by subtle violet LEDs and dozens of screens. I'm reminded of the footage from Apollo 11's mission control room on that momentous July day in 1969. The sense of excitement and discovery is palpable. To understand *OceanXplorer*'s capabilities is one thing, but to witness it firsthand is another. It inspires intense curiosity, not only at the important work being done, but at a higher level, at humanity's ability to expand its own knowledge by pushing into the unknown.

Left: *OceanXplorer* explores an atoll in the Red Sea.

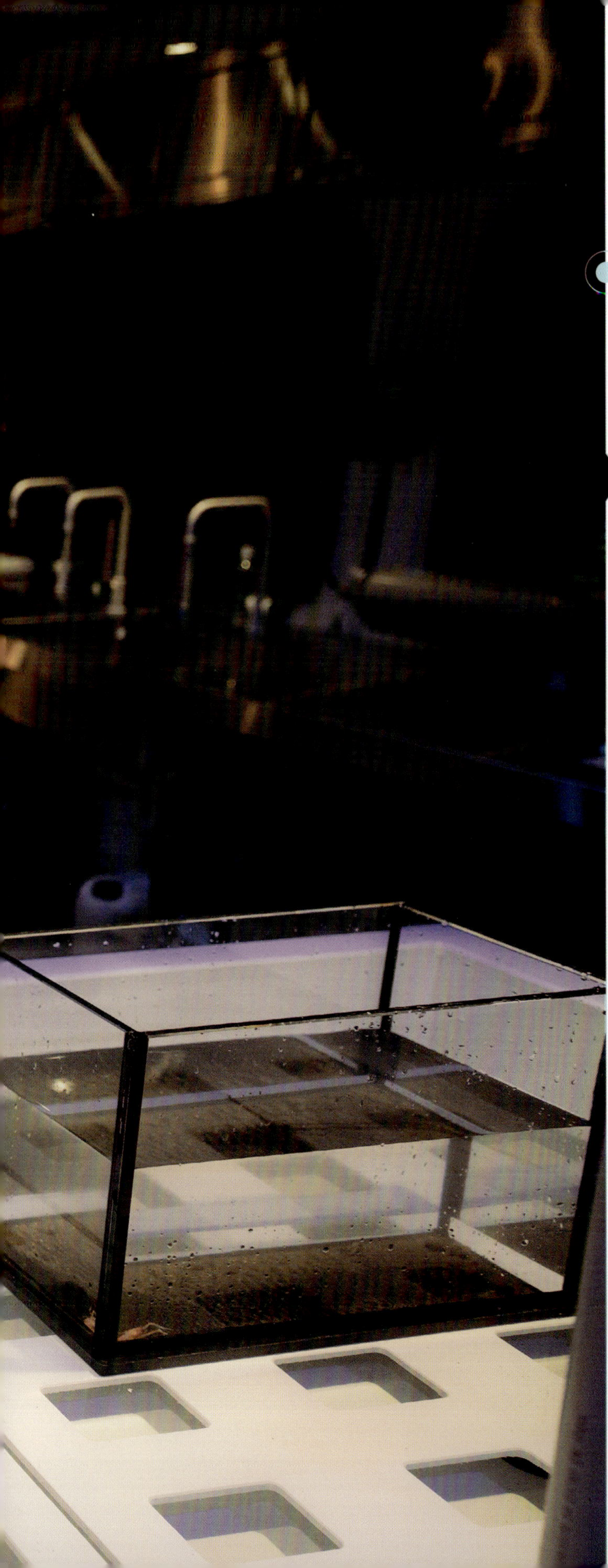

UNDERSTANDING THE OCEANS

"Yes, and?"

These are the first words Mattie Rodrigue, OceanX's science program director, uses when briefing new scientists before their first *OceanXplorer* mission. "We're always keeping the big picture in mind and coming at it through the lens of possibility," she says. *OceanXplorer* is equipped more like a land-based research institution than a research vessel, and its capabilities always surprise visiting scientists. "Say someone wants to tag whales. Well, we can also find out what's going on around the whale. We can codesign an adaptive sampling model so the tag informs our survey plan for the acoustics, and we can map what the whale is eating through acoustics, and then we can sample the water for DNA," she says. "It's big picture."

It is Rodrigue's responsibility to coordinate with the team and visiting scientists to make the most of each mission. She is astute and determined, yet warm; a bundle of positive energy that migrates throughout each department. Raised in Arizona, Rodrigue had little proximity to the ocean, but a school field trip to Scripps Pier in San Diego at the age of eleven changed it all. "My teacher threw a bucket over the side of the boat and grabbed a bunch of invertebrates for us to look at," she recalls. "I remember looking at them, and that was it; I knew this is what I wanted to do." From running fisheries programs, to mapping for NOAA, to developing diplomatic efforts to increase ocean observations, Rodrigue's passion and credentials brought her first to *Alucia* and then to *OceanXplorer*. "Our program sets out not just to change the way that people view the ocean, but to change the way that people study the ocean, and to really expand what people think is possible," she says. "That is our whole goal. That is my whole goal."

Left: OceanX Science Program Director Mattie Rodrigue inspects a sample in the wet lab on *OceanXplorer*.

Pages 84–85: OceanX crew members using the advanced equipment in *OceanXplorer*'s dry lab.

LEICA KL2500 LED

WET AND DRY LABS

OceanXplorer boasts one wet lab and three dry labs, which mirror those of a state-of-the-art laboratory at leading research institutions. They provide aquarium tanks and have capabilities that include microscopy, genetic sequencing, biofluorescent imaging, and the visualization of samples from the vessel's various sensors.

"Our program sets out not just to change the way that people view the ocean, but to change the way that people study the ocean, and to really expand what people think is possible. That is our whole goal. That is my whole goal."

—MATTIE RODRIGUE, science program director for OceanX

The processing of samples and data begins in the wet lab. "This is the busiest room by far," says Rodrigue. "Water samples come directly from the CTD and ROV next door, and the subs come in right in front of the lab." Within, three water tables are pumped with clean seawater, which is analyzed and adjusted to match the properties of the water the samples are collected in. "It's super important because we want to keep things alive long-term. We can collect deep-sea corrals from 800 meters (2,625 feet) and keep them alive for weeks just by understanding the properties of the water." This is just one of the benefits of the vessel's underway sensors, which constantly collect data and package it in an easily comprehensible way so visiting scientists can access the ship's full spectrum of data at any given time. These systems connect to a vessel-wide server that also includes information on the calibration of instruments, mapping data, submersible dive logs, and associated video files. At any moment, the team can access all the data they've amassed in real time.

Right: OceanX Science Program Director Mattie Rodrigue in one of the dry labs onboard *OceanXplorer*. These labs are packed with advanced technology to further scientific understanding and collaboration.

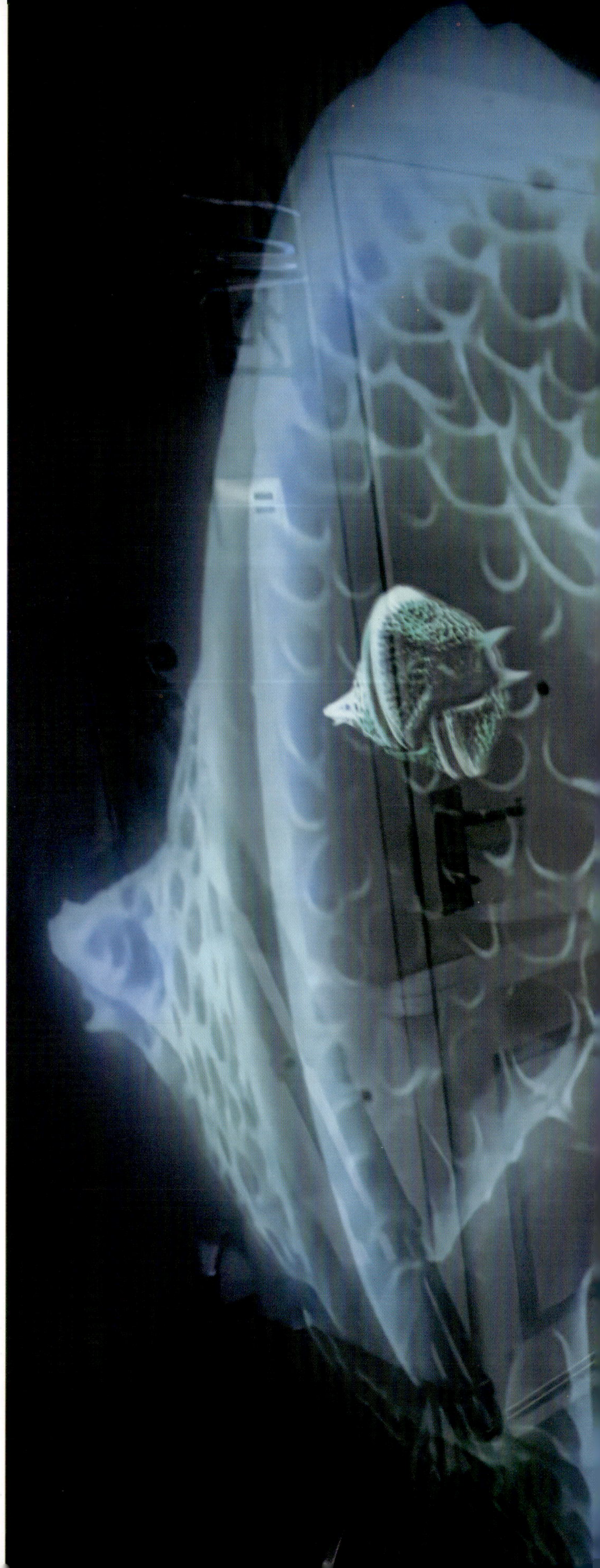

Adjoining the wet lab, the aquariums in the cold tank room can operate in total darkness to replicate the deep ocean environment, allowing for comprehensive study of the creatures found there. The labs are also curated with storytelling in mind. Camera rigs line the ceilings, and the lighting is customizable, even allowing the team to film bioluminescence. It's an ingenious feat of engineering, a perfect symbiosis between science and media.

Samples from the wet lab are then taken to one of three dry labs for imaging, sequencing, and other molecular-scale work. "Dry lab two is for imaging and microscopy, where everything is designed to be visually immersive," says Rodrigue. As one scientist studies a sample through the microscope, it is simultaneously projected onto a wall so others can study and interact with it too. Though outside scientists are frequently invited onboard, OceanX has its own research goals. The stereo microscope, for example, has fluorescent capabilities for studying biofluorescent proteins from certain animals, such as sponges or coral.

Right: Deep-sea fauna observed on a submersible dive during Mission Indonesia 2024.

Left: Navigation monitors aboard *OceanXplorer*.

Right: A large school of fish swims around a colorful coral reef in Raja Ampat, Indonesia.

Connectivity is key, and each computer is linked to a KVM system, which not only broadcasts live data or video streams from the submersibles and ROVs to the labs, but also to screens around the entire ship, so that every person onboard can witness and engage with research in real time. The list of tools to aid exploration and research is seemingly endless. A whale infrared AI system constantly searches for whale spouts, and when spotted, determines the species and logs its location. A 3D scanner scans specimens to a millimeter resolution and stores the scans in a virtual repository. A bioinformatics database holds all the publicly available genetic repositories of marine life, to which the team adds as many new species as they can. An onboard data archive and Q-GIS system holds everything they've ever geospatially documented in one local repository that onboard scientists can access at any moment. From the shallows down to 8,000 meters (26,246 feet), which is the comfortable working limit of the vessel's deepest sonar, acoustic systems on the vessel and smaller tenders work together to create one unified, cohesive map of what the habitat looks like. By developing AI systems to analyze data layers, the team can also create associations between environmental parameters, habitat, and species presence and absence, and then layer molecular information onto the map.

All this technology gives scientists aboard *OceanXplorer* the chance to fully characterize important ecosystems, such as the deep scattering layer, an area of the water column ranging from 300 to 500 meters (984 to 1,640 feet) below the surface with a high concentration of marine organisms. "It's the living connective tissue between the deep and the shallow and we're trying to understand why these organisms are so key in the trophic system and how they're really the energy that fuels the entire water column, especially for some of the larger animals in the mesopelagic area," says Rodrigue. "For me, the existential crises are climate change, deep-sea mining, and extraction of the mesopelagic area because these are all things that are so intrinsically linked to the health of the whole ecosystem. Those are the main things that we also focus on here as a program."

The most advanced technology is found in dry lab 1, the molecular lab. "This space is designed to have scientists constantly pulling DNA out of different mediums and then purifying that DNA and sequencing it," says Rodrigue. "We have two main questions everywhere we go. One, what is the biodiversity like? And two, how has that biodiversity changed over time?" All of this is achieved with the Oxford Nanopore GridION genetic sequencer and the Oxford Nanopore PromethION sequencer. Both are small and unassuming, yet they are the most complicated and powerful devices onboard.

Taken together, all of the technology onboard not only helps capture more data than ever before, but also curates the data to achieve time-saving ease of use and to advance knowledge. As Rodrigue notes, "We thought we knew what we were going to be able to do when we launched the ship, but it's so much bigger and better and bolder than what even we could have imagined and drawn up on paper."

Above: A pygmy seahorse holds onto a coral reef in Raja Ampat, Indonesia.

Right: An OceanX submersible captures biodiversity in Raja Ampat, Indonesia.

MISSION CONTROL

Mission control is *OceanXplorer*'s central nervous system, where all underwater operations are controlled. All the data collected from the vessel's deep-sea vehicles and sensors are input into a centralized mission data management platform that can be visualized in real time, both in mission control as well as on the other visual interfaces located around the ship. Outputs from the data management system can also be input into specific software for more in-depth data analysis, such as inspection of a bathymetric feature or 3D visualization of a hydrothermal vent.

This spread: Scientists assess ocean floor bathymetry in Mission Control.

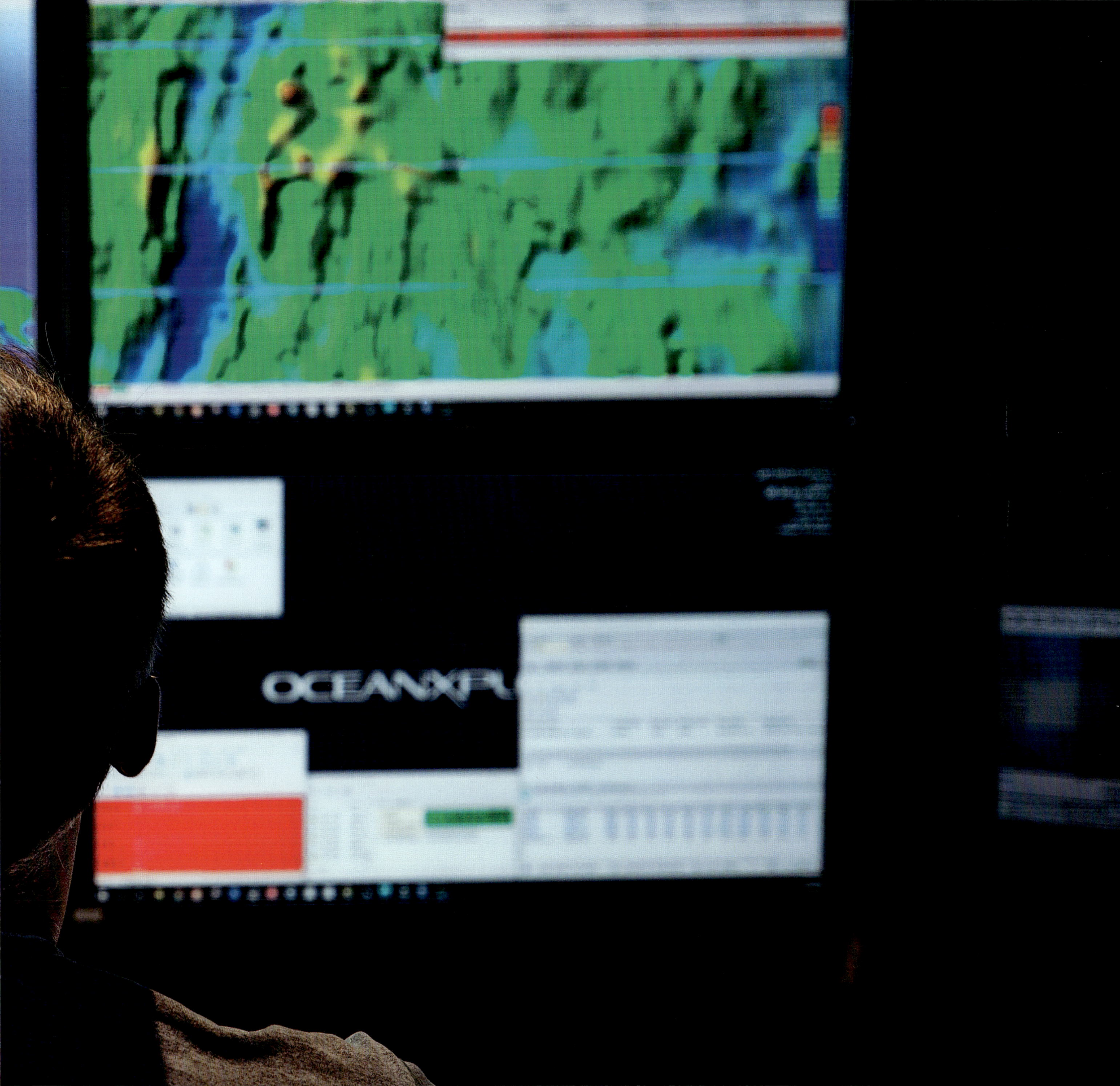

For Rodrigue, who has been onboard for every expedition to date, each mission is a treasure trove of wonders and memories, but one stands out above them all. It was a warm June evening in Exuma Sound, a deep body of water in the Bahamas. The clouds had broken open, and rain poured onto the otherwise calm sea. The team was on a mission to document and tag bluntnose sixgill sharks in their natural habitat. An illusive, prehistoric shark, the sixgill has lived in the ocean twilight zone—a layer of water that extends from 200 to 1,000 meters (650 to 3,300 feet) below the surface—since dinosaurs roamed the Earth. Indeed, teeth from a sixgill in the fossil record have been dated to 180 million years ago. Although sixgills had been tagged before, it was only possible by hauling the sharks to the surface. As a result of the trauma caused to the sharks, the behavior these surface-planted tags could document was unnatural. And so, Dr. Dean Grubbs of Florida State University's Coastal and Marine Laboratory devised a plan to tag them in the most unobtrusive way possible—by shooting a dart from a submersible at depths up to 700 meters (2,297 feet).

After two dives that failed to tag a sixgill, Rodrigue remembers standing nervously in mission control. "It was the last dive, and they only had two shots with the dart. For six hours we were waiting in silence, the tension was so high," she says. "And then it happened. The biggest moment for me was when the shark came in and it looked at the camera with its big blueish-green eye. I really got a sense of 'Oh my God. We know nothing.' We have been here for a minute compared to these creatures and we have just scratched the surface." As she speaks, I hear the wonder of the eleven-year-old Rodrigue fascinated by invertebrates found in a bucket. The wonder, novelty, the curiosity: this is what drives the team on *OceanXplorer*. "You can dream and achieve whatever you want," says Rodrigue. "I get to work with the most brilliant people on the planet. Every single time, every scientist that comes onboard walks away with this hugely expanded perspective on what can be achieved in the space that we're working in."

Left: OceanX submersible captures biodiversity in Raja Ampat, Indonesia.

"I was about ten years old when I picked up a dusty copy of Cousteau's *The Silent World* at a garage sale," says Dr. Sam Purkis, professor and chair of the Department of Marine Geosciences at the University of Miami and visiting *OceanXplorer* scientist. "I was infatuated with it, and I remember it so vividly. That was *the* moment that brought me into marine science." And though Purkis was raised far from the ocean in England's Midlands, he started diving at the age of fourteen. When he left high school at eighteen, he traveled to Egypt to work as a diving instructor. He hitchhiked across Egypt to the Sudanese border, where he lived in a tent near the Red Sea. Drinking water was sourced from a militarized zone, and wild dogs roamed the camp at night. "It was this harsh, inhospitable landscape, which then dropped into these amazing coral reefs," he says, his eyes ablaze with the memory. "It was frontier living. It was true adventure."

It was here that Purkis saw firsthand how the marine ecosystems of the Red Sea were threatened. The 1997–1998 El Niño event, one of the most powerful on record, led to the bleaching of coral reefs globally, including those in the Red Sea. Witnessing this change led Purkis to begin asking questions, to try to understand what was happening in the place he'd come to love so dearly. These questions led him back to the United Kingdom for a degree in marine biology. "I'd always go back to the Red Sea to dive, and I've had a strong connection to it for the past twenty-five years," he says. Due to these early encounters, Purkis specialized in both modern and ancient coral reef systems and has become a global authority on them by straddling marine biology and geology. "Typically, people think of marine biology and geology as separate things, but I realized early in my professional career that all of Earth's history is written in the oceans. Geology is just marine biology through time."

Pages 98–99: A blacktip reef shark swims by divers in Raja Ampat, Indonesia.

Left: Submersible gets an up close look at a coral reef in Raja Ampat, Indonesia.

In 2020, OceanX announced *OceanXplorer*'s maiden voyage to the Red Sea with a simple question in mind: How do species in the Red Sea thrive in conditions in which others die? A team was amassed, and Purkis was invited to join. With expert knowledge in the area accumulated over many years, Purkis was in a unique position to guide the team and explore the areas he believed held new findings. "I was interested in the formation of the Red Sea. I'd only been down as far as a scuba tank could take me, but now we were going to go down into the abyss," says Purkis. "The ship is awe-inspiring, with the submarines and the ROVs. It's incredible, but the real power is beneath the waterline with its incredible multibeam [sonar] system, which maps the seafloor." From mission control, Purkis watched as the data came in in real time. Anticipating it to be barren, he gleaned a view of the seafloor replete with mountain peaks, canyons, and trenches, and he knew they'd be able to make remarkable new discoveries down there. The submersibles were readied, and Purkis began his first dive.

"The hatch was closed, there were neon lights, people running around plugging in gas bottles. It felt like being an astronaut getting ready to be launched into space," he recalls. "And then it was almost impossible to process that we were descending into the abyss, and within seconds we were beyond the deepest I had ever been in the ocean. It's the thrill of exploration and the thrill of discovery. You're visiting a place on Earth that no one's been to before, and you don't know what you're going to find."

Above: A Triton submersible aboard *OceanXplorer*.

Right: The crew of *OceanXplorer* prepares to launch a submersible.

This spread: An OceanX submersible among hammerhead sharks. Photo by Jonathan R Green.

At 900 meters (2,953 feet) below the surface, Purkis made his first discovery, witnessing clearly that the seafloor was sliding into the central part of the Red Sea. It was buckled, fractured, and splintered, composed of material that should have been found in much shallower locations. It confirmed his hypothesis that the whole seafloor was on a conveyor belt and moving, and it's from these observations that further research has begun into understanding how oceans develop over time. Purkis explains that the Red Sea is a baby ocean, much like the Atlantic would have been 170 million years ago, and it holds a wealth of information to further our understanding of how oceans are formed.

While Purkis was observing the seafloor from the submersible, fellow scientist Dr. Francesca Benzoni, a professor at the King Abdullah University of Science and Technology, was documenting life. "She was finding organisms she had never seen before. She was fizzing with excitement, and it made the whole experience overwhelming in the best way," recalls Purkis with a wide smile. It's this ability to bring scientists together in cross-disciplinary collaboration that sets OceanX apart."

“Curiosity and adventure are what drives us today. The thrill of discovery is addictive.”

—**SAM PURKIS**, professor and chair of the Department of Marine Geosciences at the University of Miami and visiting *OceanXplorer* scientist

In the Gulf of Aqaba at the northern extent of the Red Sea, Purkis analyzed the data from the sonar array and discovered an area over a mile deep. From the faults and fractures, he believed the elusive brine pools might exist there. Others had been found before, but never in this location. A mission was launched with the Argus ROV to descend to more than 1,500 meters (4,921 feet) to explore the seafloor. “It was just a muddy desert, not even a pimple or a rock,” says Purkis. “One hour passed, two hours, seven hours, and there was nothing.” After ten hours, Purkis was deflated and concerned that he'd wasted the day, the team's time, and precious resources. He made the call to retrieve the ROV. Some missions don't always lead to success. However, sensing Purkis' dejection, ROV pilot Ewan Bason told him that they could push on for another ten minutes. “And then there was a blur, a weird haze in the video feed. The water was almost like gel, but it was a brine pool,” says Purkis. “It was absolutely spectacular, an alien landscape with all sorts of weird organisms and microbes. Suddenly, all the scientists and crew were crammed into mission control and huddled around the video.”

Right: An OceanX submersible surveys a sunken airplane propeller at Bikini Atoll in the Marshall Islands.

OCEANX
OCEANXPLORER

"Every detail on *OceanXplorer* is designed to facilitate visually compelling storytelling about marine science."

—EBEN HALL, executive producer for OceanX

SONAR ARRAY

The sonar array (known as the gondola) features two multibeam systems, the Kongsberg EM712 and EM304, capable of mapping to depths of 3,600 meters (11,811 feet) and 6,000 meters (19,685 feet) respectively. This acoustic array enables high-resolution mapping and exploration of the seabed, also known as bathymetry. Working alongside other sensors, including those that visualize the top layers of seafloor substrate, currents, and biomass within the water column, these technologies provide scientists with all the data they need to deeply understand an entire ocean ecosystem from surface to beneath the seafloor.

Left: *OceanXplorer* is built to explore the deep ocean, like these Norwegian fjords.

The pool they'd found extended to almost 10,000 square meters (107,639 square feet), with smaller pools dotted around the fringes. Because brine pools are anoxic (lacking oxygen) and unable to sustain animal life, there is no bioturbation (reworking of the sediment by flora or fauna) on the seafloor, meaning that core samples from beneath the pool could shed light on the climate of the Red Sea throughout history. "The bottom of the core sample was 2,000 years old. We found evidence of earthquakes, tsunamis, heavy rainfall, droughts, everything that has been going on in that area for the last 2,000 years," says Purkis. "It's almost as if you put instrumentation on the seafloor to record these things and left it there for millennia. That record is more exquisite than we could ever imagine." It's a discovery that they will be working on for years to come, and its results will further our understanding of the ocean.

Purkis was astonished at how swiftly the team was able to build a compelling story around the discovery. "It's that ability to react quickly to discoveries, and an incredible blend of media and exploration that sets OceanX apart," says Purkis. "This is what Cousteau did, and OceanX has taken that idea and run with it much farther than could ever be imagined." The reach of this surprised Purkis, who recalls receiving messages sent by parents from his daughter's school and friends he hadn't spoken with for twenty years with words of awe over the video OceanX shared about the brine pool discovery. "Media is as important as the science, and science is small and silly without storytelling; there's no point doing any of this if the message is not transmitted out," he says. "OceanX is very nimble at focusing on the positives, and there are many positives. It isn't all doom and gloom. It's super important to transmit that to young people. I changed my entire career through contact with a dusty old book, and I'm sure that on a daily basis OceanX is doing the same to young minds."

Right: Submersible shines its lights on a swarm of krill in the deep ocean surrounding Antarctica.

Right: *OceanXplorer* utilizes a suite of assets to explore marine environments overhead and up close.

Pages 114–115: The sub hangar aboard *OceanXplorer* houses all submersibles.

HAPPY BIRTHDAY DEAR E
TRITON

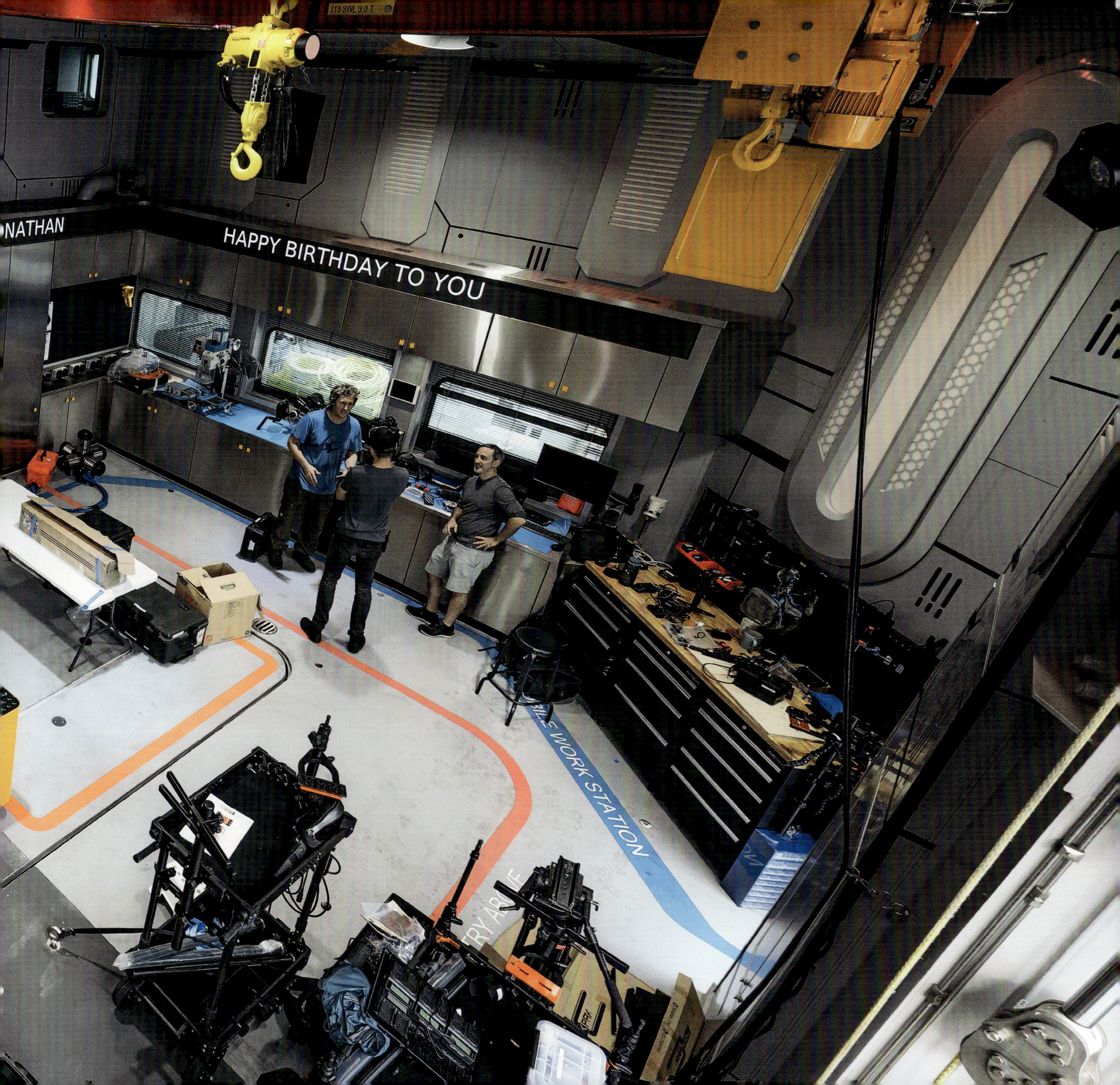
NATHAN
HAPPY BIRTHDAY TO YOU
WORK STATION

evotec
OCEANX
OCEANX

THE WINDOW TO A NEW WORLD

"We're the candy in the sweet shop. Our scientists and collaborators come to this boat, and what we can do for them is only limited by their desire for research," says Andrew Craig, *OceanXplorer*'s ROV team leader. "We've really strived to make sure that the team around this technology are at the top of their game and can work together to get the most out of it."

Born into a family of farmers, Craig specialized in heavy agricultural machinery and had little connection to the oceans. However, on a journey to Australia, that all changed. "My family are Irish farmers—there are no seafarers—but I started to dive in Australia, and my career path took a completely different direction from that point on," he says. "I loved the adventure and the exploration. Being underwater was one thing, but also taking the engineering and technology side and seeing how that's applied underwater really interested me." This led him to a degree in marine technology, then into archaeology, and a twenty-year career in deep-water shipwreck exploration and recoveries. But with *OceanXplorer*, he's found a greater purpose. "The whole thing just feels right," he says, "especially when you start a mission with these completely different scientists who you've never met before, and by the end of it they're crying or they're hugging you. They get really engaged in the boat, and to see them go away with a dataset that would normally take them years to get is very fulfilling."

Left: OceanX crew members inspect the ROV aboard *OceanXplorer*.

Right: *OceanXplorer* launches its CTD and ROV in the fjords of Norway.

CHIMAERA: ARGUS MARINER XL ROV

Tethered to a cable as it descends thousands of feet below the ship, Chimaera illuminates the deepest reaches of the ocean. Its high-definition, 360-degree VR camera and lighting system capture rarely seen creatures in stunning quality while fiber optics in the cable connecting the ROV to mission control enable it to relay footage in real time.

Mariner
XL
OCEANX

> “The ocean is not just this far- fetched dream that somebody else owns; it’s a resource that should be accessible to everyone for education and inspiration.”
>
> —**MARIO TADINAC**, media technology officer aboard *OceanXplorer*

Operating the ROV is a three-person task. As the ROV is powered by and tethered to the vessel with a 6,500-meter (21,325 foot) cable, one person controls the launch and recovery system (LARS) to ensure a seamless mission. Just meters away in mission control, a pilot and co-pilot sit before a bank of screens linked to the ROV’s fourteen cameras. One pilots the ROV, and the other controls the arms, which can lift up to 120 kilograms (265 pounds) at full extension. A remarkable robotic arm in mission control mimics motion of the arms on the ROV, allowing the team to collect an array of samples. Tools and techniques used for sample collection include pushing tubes into sediment for core samples, utilizing the suction to pull organisms or samples into a carousel with eight individual chambers, and a bio-box for larger samples and organisms. It’s delicate, skilled work that only succeeds through expert collaboration. This extends to the bridge, where the captains control the vessel’s dynamic positioning so that the ship moves in unison with the ROV.

Right: OceanX’s submersibles, *Nadir* and *Neptune*, descend in the waters off of Norway.

SHOGUN

"The ocean is like a giant toy for scientists and explorers, but it's a toy that holds the key to our planet's future."

—**ANDREW CRAIG**, ROV team leader aboard *OceanXplorer*

The ROV itself is equipped with scaling lasers to take the measurements of creatures and features at depth. In addition, two sensors record position acoustically, and a multibeam sonar system paints high-resolution images of the seabed. The state-of-the-art inertial navigation system (INS) is the most advanced piece of technology on the ROV, and allows the team to know where the ROV is to within 10-centimeters (3.9 inches). Along with the ROV's exploratory functions, it is also designed with storytelling in mind. Booms extend from the ROV to allow for additional cameras and lighting. The 120,000 lumen LEDs are all dimmable, and a 700-nanometer red light allows the team to observe creatures of the deep without disturbing them. Two wide-angle cameras are part of the 360-degree VR camera rig, which allows the team to watch through VR goggles. It's camera technology like this that has changed the very nature of observation at depth. "In the last 20 years, it's gone from being a grainy, noisy black-and-white single image to four 8k cameras with full automation," says Craig. "The engineering and technology have also gotten so advanced and reliable that the ROV can stay down for days and days and days."

Left: An OceanX submersible surfaces during a mission in Antarctica. Photo by Paul Nicklen.

All of this technology allows for exploring and documenting the deepest parts of the oceans in ways unimaginable just a few years ago, and it's this that Craig notes as the greatest success. He becomes animated as he talks about the maiden voyage to the Red Sea. "They're calling the coral in the Red Sea the 'canary in the coal mine' because the Red Sea is so hot. If they can figure out why the coral in the Red Sea is so healthy, they can begin to understand what's going on as waters in other oceans are heating up," he says. "We worked extensively with the scientists who found ten new species, which is super interesting and a direct result of us helping that research." On another mission in the Red Sea, they discovered Humbolt squid around an unexplored shipwreck. "Nobody ever thought they'd find them in the Red Sea, and we saw them for the first time with the ROV," he says. "The scientists stand behind us in mission control and see everything live, and seeing their reaction to what we're seeing is exactly why we do what we do."

Right: A close up of a marine speciman sampled aboard *OceanXplorer*.

OCEANX

OTTER UNMANNED SURFACE VESSEL (USV)

The Otter is an autonomous surface vehicle designed to go to areas too dangerous or too shallow for *OceanXplorer* to maneuver. It has a full suite of mapping sonars for the seafloor, water column, and even shallow reef systems, as well as a 360-degree camera, LiDAR, and a hyperspectral camera.

This spread: OceanX's Otter USV maps a shallow area in the fjords of Norway.

Pages 128–129: A flourishing coral reef within the Santa Maria marine protected area of the Azores.

NADIR
T/T ALUCIA

While the ROVs offer unmanned exploration below 1,000 meters (3,280 feet), much of the exploratory magic is achieved with the submersibles. Human operated, with space for a pilot and two passengers, *Nadir* and *Neptune* tap into the curiosity and thrill of adventure that is so integral to the OceanX philosophy. "What we do is a total sensory overload at times," says David Pollock, *OceanXplorer*'s submersible team leader. "Nothing can prepare you for what you witness and experience."

Pollock grew up in the county of Fife, Scotland. "As a kid, I always loved water and going to the seaside and peering into rockpools," he recalls. "I started working on navy ships when I left college at seventeen and then in defense engineering for a number of years before working on submarine rescue systems, and I was hooked from then on." While he honed his skills building, maintaining, and piloting rescue submarines off the coast of the United Kingdom, *OceanXplorer* offered a new way to utilize his skills. "I've always been interested in documentary filmmaking, media, and science," he says. "It's so important, and that's why I do it."

Left: The OceanX submersible *Nadir* launches in Raja Ampat, Indonesia.

> "The sense of adventure and curiosity in marine science is what keeps us exploring the unknown."
>
> —**STEPHEN HUDSON**, dive safety officer aboard *OceanXplorer*

Pollock worked closely with Triton to modify the two Triton 3300/3 submersibles for the needs of *OceanXplorer*. *Neptune* is equipped for scientific research and includes bio boxes for larger sample collection, water suction devices for smaller organisms, and core drills for samples of the seabed. "Anything the scientists can think of, we've got something to be able to sample it and bring it back to the surface so they can study it in greater detail." *Nadir* is outfitted to enable media documentation of the highest level and includes cinema-quality cameras with interchangeable lenses to adapt to all conditions. "Ninety-five percent of the time, both subs go in together," says Pollock. "Depending on the mission objectives, they may be a few hundred meters apart or working within a few inches of each other during filming."

When up close to *Nadir* and *Neptune*, there's something almost mythological about them. Their vast 16.51-centimeter (6.5-inch)-thick acrylic domes are like the cyclopic eye of some deep-sea creature; their ballasts and thrusters are like a carapace, with mighty decapod-like claws that reach forward. And while humans have created them, they are indeed creatures of the deep—the only machines that allow us to obtain an unobscured vision of the dark underwater realms with our own eyes. They have enabled an array of discoveries that would otherwise remain folklore.

Right: *OceanXplorer* with its submersible and ROV on display in the fjords of Norway.

OCEANXPLORER
VALLETTA

NADIR AND NEPTUNE

Descending 1,000 meters (3,280 feet) into the twilight zone, the ship's two manned Triton submersibles are equipped with an array of science and media equipment. *Neptune* is configured to maximize research with several scientific mounting and collection devices. *Nadir* doubles as a mobile marine studio, broadcasting its discoveries to the surface in real time and allowing storytellers to document the deep in high-resolution.

This spread: OceanX crew member unhooks one of the submersibles from the vessel before a dive.

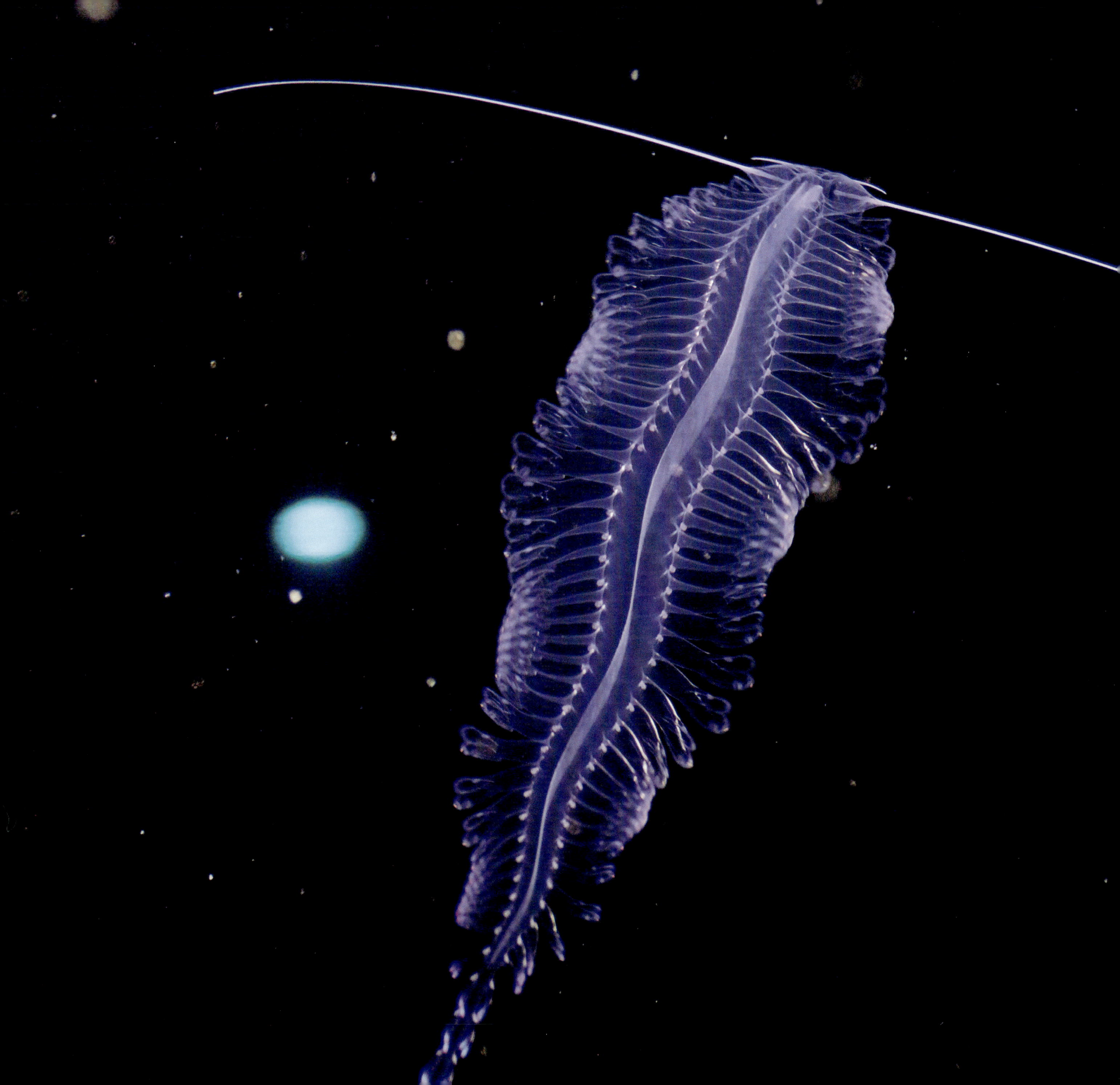

On a calm June day in 2021, the submersibles stand ready to enter the water. In the ocean surrounding the Azores, the team has been searching for bioluminescence in the deep scattering layer at approximately 500 meters (1,640 feet). Lifted by the ship's A-frame and released to dive, *Neptune* and *Nadir* take thirty minutes to descend into the depths. To preserve any potential bioluminescence, they go in dark: Every LED switch within is taped over, and the control panel is covered. Pollock peeks at it regularly to see where they are. By going in dark and then flashing the submersibles' lights, a chemical reaction occurs in any nearby bioluminescent organisms, provoking them to flash back. As they enter the deep scattering layer, Pollock puts *Neptune* in neutral. With the signal given, he momentarily flashes the lights. "You know about these things, and you know they exist because you've worked with scientists and you've read books, but nothing can prepare you for actually experiencing and witnessing it," he says. "You're not expecting it. You flash the lights, and the whole place around you lights up with blue flashes. It's just unbelievable. It's like all the stars suddenly flashing at you." For up to 100 meters (328 feet) around the submersible, the bioluminescence appears in small dots and larger thread-like structures. We still don't fully understand why this occurs. Is it an alarm call? Is it a form of communication? Due to the difficulty in accessing the deeper twilight zone, observation and documentation of this phenomenon is still in its infancy. It's only by embarking on such missions that our understanding will improve.

While all the action today is around the submersibles, nothing happens on *OceanXplorer* without vessel-wide collaboration. "There's always a comparison to a family with the ship's crew," notes Pollock. "Everyone is really focused, really driven to help and achieve what we set out to achieve. We have very happy scientists, very happy media crew, and that's a testament to all the ship's crew as we rely so heavily on one another."

Left: Close up of a gossamer worm filmed from OceanX's submersible in Indonesia.

Beside the sub hanger, a staircase leads down to the dive center. While deep-sea exploration is critical to furthering our understanding of the oceans, much of the ocean's megafauna and other marine organisms are observed in the epipelagic zone, also known as the sunlight zone, which extends from the ocean's surface to about 200 meters (656 feet) down. In this sunlight zone, where most of the visible light in the ocean exists, there is much that thrives in the upper 50 meters (164 feet), and it is easily accessible to divers. The dive center was designed by Stephen Hudson, *OceanXplorer*'s dive safety officer, to maximize access to this zone and minimize the time it takes to deploy divers. "I learned to dive at an early age and became an instructor at twenty. I've now been running dive expeditions for twenty-plus years, hence the gray hairs," he chuckles. "I designed these spaces with flow and practicality in mind. To see them change from old, dusty storerooms to spaces used by divers is a great privilege. I wanted it compartmentalized, so we could have a dry office, an area for equipment storage, and a photo- and video-editing suite. With separate dedicated wet rooms, including a dive lobby for changing into gear and a dive wet camera workshop, we have space for all our pre- and post-dive needs."

These rooms allow direct access from the ship's side boarding platform for each loading and unloading from the tenders. Also in this area is the ship's onboard hyperbaric chamber, in the unlikely event that any divers experience decompression sickness. This is a rare yet highly valuable piece of safety equipment to find on a research vessel, and fortunately it has never been used in an emergency. Beside the hyperbaric chamber is a cavernous gas storage room that holds a six-to-twelve-month supply of oxygen and helium for both the dive and submersible teams. Although Hudson primarily runs the dive center and operates as the dive safety officer for visiting science and media teams, he's also a skilled camera operator and often assists with filming assignments in more challenging dive environments.

Right: OceanX crew members stand on the aft deck as the submersible launches from *OceanXplorer*.

TRITON
OCEANX
MORELLO

OCEANX

OCEANX

DIVE CENTER

The fully equipped dive center allows scientists and storytellers to be dropped into untouched habitats in minutes using the vessel's rapid-deploy Zodiacs, rigid inflatable boat (RIB), and custom 38-foot Metal Shark research tender. While in the water, the dive team collects samples, evaluates marine life, and documents the environment using state-of-the-art underwater camera rigs.

This spread (clockwise from top left): The dive center on *OceanXplorer* is fully equipped with scuba diving equipment, allowing scientists and storytellers to get up close with marine environments; Scuba divers observing jellyfish in Raja Ampat, Indonesia; *OceanXplorer*'s dive center is also equipped with a hyperbaric chamber.

It's March 2021, off the Navidad Bank north of the Dominican Republic. Hudson and the filming team have been following four humpback whales when he receives a call from the helicopter team. Ten nautical miles out to sea, they have just spotted a large pod of orcas hunting a female humpback and her newborn calf. "It was a long way on our small boat, and the sea conditions were very choppy, but it was a once-in-a-lifetime opportunity, and we got there as fast as we possibly could," he recalls. Working with cetaceans is a fine art; a respectable distance must be maintained to avoid disturbance, but the team must also be positioned close enough to document these fast-moving animals. While the helicopter films from above and scientists document from the other tender boat, the dive team enters the water to observe below the surface. "Watching a pod of more than a dozen killer whales trying to separate this young calf from its mother was quite emotionally draining," says Hudson. "After two and a half hours, it was quite upsetting to see four orcas drag this calf into the depths, but it was a huge privilege to be in the water, because its generally not ever something you would capture on camera."

Right: A scuba diver surveys a coral reef in the Red Sea.

Later that same day, the enchantment of the ocean erupts when the team encounters another female humpback with a four-week-old calf. "It was so wonderful. We got in the water, and the calf was incredibly playful and inquisitive," says Hudson with a wide smile. "We took it very slowly, and the mother was very accommodating. The calf was so inquisitive that it came to within a meter of me." All the while, Hudson filmed the encounter. "When you see these animals and the environment, it's so beautiful."

Only by witnessing and documenting such encounters can we further our knowledge of cetacean behavior. Helping capture these moments is what drives Hudson. "If you're a nondiver, someone who will only ever see it on a television screen, the message is incredibly important, because otherwise how do you appreciate it?" he says. "If you're able to see it, then there's a valid reason for why you need to be really conscious about what you do on land, because it affects the ocean." While research in and of itself is incredibly important for broadening our understanding of the oceans, the message only reaches a truly global audience through the power of storytelling.

This spread: Clownfish hides in an anemone in the waters near Solomon Islands and Papua New Guinea.

Pages 146–147: A pod of dolphins break surface in Raja Ampat, Indonesia.

THE POWER OF STORYTELLING

This spread: *OceanXplorer* at sea off the Azores.

"Science is the key to saving the world. To protect something, you have to understand it," says Eben Hall, OceanX's executive producer. "For that science to happen, the broader public needs to be excited and inspired. That's why capturing the science and packaging it in a way that is informational and inspiring is so important."

OCEANXPLORER

Hall grew up in the deep woods of Maine, where he roamed the wilderness and spent his youth sailing with his father. "I've always felt very connected to nature and comfortable on the ocean," he says. "It's always been a part of my life." Early in his filmmaking career, he went on an assignment to Haiti on the one-year anniversary of the catastrophic 2010 earthquake. It was both the adventure and the importance of documenting such a pertinent story that gripped him and led him to pursue stories that combined both. "I find documentaries and storytelling so satisfying, because you get to experience amazing places and people and then capture it all on film to share with the world." While creating short-form documentaries for CNN's *Great Big Story*, Hall began to focus on wildlife conservation, which caught the attention of Mark Dalio.

In spring 2021, Hall travelled to the Azores and Norway to follow *OceanXplorer* and produce short documentaries about local scientists. Gripped by the vessel's media capabilities and the important research conducted onboard, Hall joined OceanX's in-house storytelling team. "Every detail has been designed to facilitate really visually compelling storytelling," he says. "From a technological and practical standpoint, it couldn't be more easy or pleasurable to produce content onboard." Every LED on *OceanXplorer* can be programmed to create studio-like lighting for interviews or for capturing footage; the media center is equipped with the latest editing software and file storage systems; and a dedicated wet media lab allows for seamless preparation and maintenance of underwater camera equipment. In its conception, *OceanXplorer* was created to make storytelling as intuitive as possible. "Everything is happening all in the same place. The media center is a thirty-second walk to the sub hanger, which is a ten-second walk to the ROVs and mission control," says Hall. "The big advantage of *OceanXplorer* is that it's multiplatform. The subs and ROVs might be down while the helicopter is in the air, so it's awesome to be able to pivot so quickly, especially when everything is happening at the same time."

Right: Free diver in the Azores.

MEDIA CENTER

Mixed-reality data visualization and real-time editing allows the media team to deliver short-form, episodic, and live content while at sea. The content they work with comes from cutting-edge low-light and high-resolution cameras carried in custom housings, including the only housing in the world capable of shooting 8k resolution at nearly 6,000 meters (19,685 feet) below the surface.

This spread: An OceanX submersible surveys the seafloor of the Red Sea.

OCEANXPLORER
MAJURO
OCEANXPLORER
MAJURO

It's a still, blisteringly hot day in February 2022. *OceanXplorer* lies off the coast of Saudi Arabia in the Red Sea, with Hall in mission control filming scientists planning the day's research. That's when he receives a call from the helicopter, which is conducting aerial surveys. An enormous whale shark had been spotted 6 kilometers from the vessel. "Within minutes, the RIB was in the water and we were powering out, with the helicopter overhead keeping an eye on the whale shark and guiding us in. With the support of the helicopter and being able to react so quickly, our underwater director of photography was able to get the shot," he says with animation. "It was an incredible shot. Of the thousands of hours of footage, that was probably the most beautiful thing that came out of the whole expedition. That's when I realized the power of *OceanXplorer* for natural history from a storytelling perspective."

Beneath the surface, Hall is also responsible for documenting the submersibles' deep dives. While *Neptune* performs its scientific operations, Hall and the team follow and document in *Nadir*. "When you see the surface close over you and you go down until it's pitch black, there's so much excitement and romance. Often the most interesting shot is to look at the other sub and see it in context with amazing topography," he says. "It's important to actually go and see these things with your own eyes, and it's more inspiring to people who are watching the films later."

Pages 154–155: *OceanXplorer* tows submersible for retrieval after an eventful dive.

Right: The OceanX submersible *Neptune* collects samples from an Azores seamount.

Page 158: A free diver meets face-to-face with a whale shark at Isla Mujeres, Mexico. Photo by Christian Vizl / Tandem.

Page 159: A great white shark swimming near the surface of the ocean at Guadalupe Island, Mexico. Photo by Christian Vizl / Tandem.

The proximity in which the media crew and scientists live also helps foster strong relationships, and therefore stronger storytelling. "The great thing about *OceanXplorer* is that not only are you working with scientists, but you're living with them, too," says Hall. "There's a very social element, so you quickly become friends and they become comfortable with you and the cameras. What you're doing together is so fun and cool and amazing that nobody takes it for granted."

Along with the onboard media, Hall also works with a rotating crew of filmmakers utilizing *OceanXplorer* for more expansive storytelling. The largest project to date is the TV series *OceanXplorers*, a joint production of OceanX, the BBC Studios Natural History Unit, and the National Geographic Society. This groundbreaking six-part series—which aired in August 2024 and has reached viewers around the world on the National Geographic Channel, Disney+, and Hulu—is the most ambitious ocean exploration and research series ever created, and it is the zenith of what OceanX set out to achieve. "For a general audience, I think that the exploration and the adventure is the spice that brings the science to life and makes it more palatable and inspiring," says Hall. "The fact that we are doing this amazing, important science, while at the same time having this grand adventure, makes for a powerful combination."

OCE

"My first impression of *OceanXplorer* was beyond mind-blowing, and what blew my mind was the science, the capability of the boat, but most interestingly the people," says filmmaker and *OceanXplorer*'s media technology officer Mario Tadinac. "It was the mystery and charm of *OceanXplorer*, where it brought so many explorers together and where people opened up and could be human."

Tadinac first joined *OceanXplorer* in January 2021, and he subsequently spent eight months onboard documenting all aspects of the research. As a storyteller, it is the exploration in combination with the intrigue of the people behind the research that gripped him. "So many times I would be filming scientists in the labs working with samples, but it's the journey to how this came to be, discovering people living their childhood dreams," he says. "As a filmmaker, you always look into how you can make it personal by capturing the essence of who these people are, which is ultimately the passion they have. That is the beauty that we pursue."

Left: A bull shark swimming alongside fish in Playa del Carmen, Mexico. Photo by Christian Vizl / Tandem.

Born and raised in Split, Croatia, Tadinac lived beside the Adriatic Sea, where life was intertwined with the ocean. "Being next to the water always feels like being at home," he says. "What the oceans gave me as a child, I want to give others through storytelling."

Tadinac's love of storytelling is also fused with a skilled technical mindset. Onboard, he helped curate the media center to facilitate seamless access for productions. This also includes the fine-tuning of live feeds and mixed reality to embolden greater storytelling. "It's amazing the way the sub can capture footage that's beamed straight into the media center, where we figure out how to narrate it and share it with the world live," he says. "It's also the journey of finding out what goes through the minds of scientists who are down there in places we've never been before as they are capturing it and logging it. Scientists have their own language, and it's about finding the simplification that connects the gap between them and the viewers."

Right: *OceanXplorer's* submersible illuminates the depths in the Azores.

OCEANX

HELICOPTER

From a specially designed hanger on *OceanXplorer*'s bow, the Airbus H125 helicopter is equipped for long-range search missions, deployment of research teams, and filming and documenting from the skies.

This spread: *OceanXplorer's* helicopter allows researchers to survey the ocean from the skies.

Right: *OceanXplorer* hosts a helicopter that facilitates megafauna surveys and shallow ecosystem assessments.

Pages 168–169: Participants of OceanX's Young Explorer's program jump for joy on the helipad in Indonesia.

Tadinac has been present for many of *OceanXplorer*'s missions. He recalls capturing footage of a vast underwater mountain from *Nadir* in the Red Sea, filming glaciers in Svalbard in the Arctic Ocean, tagging ancient Greenland sharks, and documenting walrus colonies. Yet one memory stands out above them all. "One of the most gratifying scenes I had the pleasure of filming was tagging sperm whales in the Azores," he recalls. "It was tedious; we were working hard and waiting for nature, but then we found them." In a rare observation caught while scouting with the drone, the team discovered four sleeping sperm whales. They approached slowly, careful not to disturb the whales. When they were close enough, marine researcher and National Geographic Explorer Rui Prieto successfully placed a suction-cup camera tag on one of the whales, which would later be released and collected from the surface.

Because sperm whales hunt for giant squid at depths of more than 2,000 meters (6,562 feet), almost no footage of this behavior has been captured before. As the whale dove, its speed was so great that the camera began to rotate and point vertically, offering no view of the whale itself. The excitement over such amazing footage soon turned to frustration. "All this effort, all these resources, and you feel the scientist's disappointment," says Tadinac. "But the silver lining is that the tag eventually fully rotated, and we were looking backwards." For Prieto, this was a chance to witness, for the first time, the tremors and vibrations of the sperm whale body as it hunted, to see the tail and flukes quiver as it predated on giant squid. "You didn't get what you really wanted, but you got something totally unexpected that's still really fruitful," says Tadinac. "It's very rare as a filmmaker. You film so many hours of pursuits, and then you're given these gratifying moments."

Not only does footage like this aid scientists, but it can be used to inform and educate a wider population. In this way, filmmaking onboard *OceanXplorer* is multipurpose. "If we are given the privilege to be on this boat, let's give that privilege back to the world so they have access to this information. *OceanXplorer* is a floating hub for scientific research, for education, and for inspiration," says Tadinac. "As a storyteller, it's about telling the stories of these people and inspiring future generations."

At the end of a long day, the crew takes a moment to unwind in the comfort of the lounge. This is a sacred space, where smiles spread on the faces before me. There are conversations, jokes, and laughter. While these are the leading experts in their respective fields, the familiarity with which they treat one another is touching to see, and I'm reminded of the crew aboard *Calypso* in the groundbreaking TV series *The Undersea World of Jacques Cousteau*, a show that inspired many of the people here. Professionals, yes, but also friends. And in this observation, something occurs to me. Although *OceanXplorer* is a vessel curated with the utmost precision and with equipment that sits right on the cutting edge of technological advancement, it means nothing without the ingenuity, curiosity, and collaboration of its crew, scientists, and storytellers. Only in unison can such a platform change the way people study, document, and communicate the wonders that lie below the surface. And that is the power of *OceanXplorer*.

OCEANX
N988MD

PART III:

IMPACT AND LEGACY

OCEANX
OCEANX
OCEANX
OCEANX

THE CATALYST OF CHANGE

OceanX is comprised of three core pillars: science, media, and education. Through these pillars, and aided by *OceanXplorer*, the team has been able to produce a remarkable global impact by inspiring youth, aiding scientists, making ground-breaking discoveries, producing award-winning media, and engaging in policy and governance dialogues aimed at both the conservation of the ocean and the prosperity of those who rely on it.

OceanX Education exists to develop the next generation of ocean leaders. It is impossible to address the complicated challenges facing the ocean and the planet without attracting a diverse and engaged group of young people to find solutions. Core to its mission is a range of education programs that provide future ocean leaders with hands-on learning experiences, including Young Explorers missions that invite students aged eighteen to twenty-two to embark on an immersive multiday excursion onboard *OceanXplorer*. Through this and other on-ship, digital, exhibition, and outreach programs, the goal is to inspire students to become ocean explorers, problem solvers, advocates, and storytellers by offering opportunities to learn about and engage with the ocean.

"OceanX is the reason why I am a marine bio major," says Madeleine Stewart, an undergraduate at the University of California, Santa Cruz, and a participant in the first Young Explorers Program in July 2021. "I've always loved the oceans; I just didn't know I could actually study them for a living. This was the first time I was exposed to marine scientists. I got to meet them and hear their stories. If I had not gone on this trip, I would not be a marine bio major. It redirected my entire life for the better."

While growing up in Washington, D.C., Stewart spent her summer holidays exploring the Sea Islands of South Carolina. "I thought being around the salt marshes, seeing dolphins and alligators, was the coolest thing. It's just so beautiful. I can't remember a time in my life where

"We really need to continue to peel back the unknown and keep learning. The greatest era of exploration is still ahead of us. Exploring the ocean can help provide a piece to the big puzzle."

—DR. SYLVIA EARLE, marine biologist and explorer

Pages 170–171: *OceanXplorer* in Dubai harbor.

Right: Young Explorers learn about the functionality and importance of CTDs during missions.

Pages 174–175: Young Explorers stand on deck of *OceanXplorer*.

FRONT
FRONT

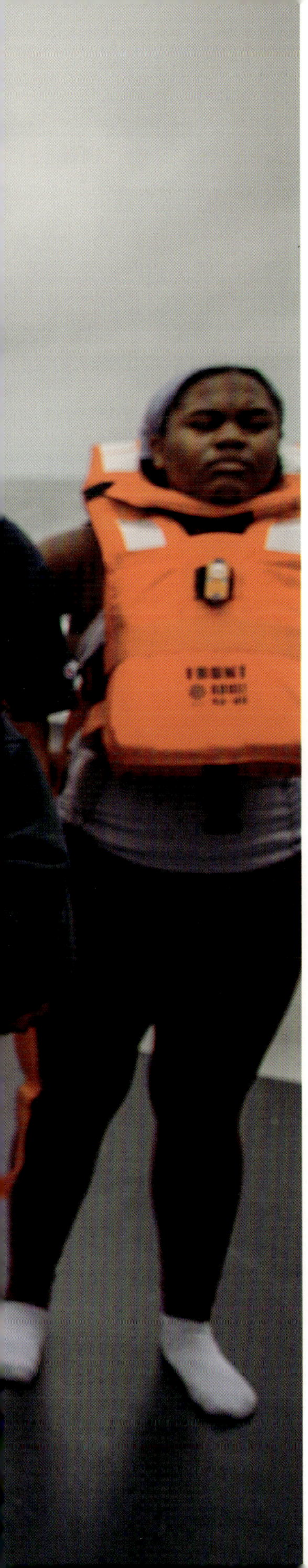

"We are on a mission to explore the ocean and bring it back to the world, and we believe this must include nurturing and educating the next generation of scientists. The Young Explorers Program is an opportunity to share our learnings, technologies, and passion for the ocean with exceptional young students and is a great step in ensuring our oceans are understood and protected for generations to come."

—**MARK DALIO**, cofounder and co-CEO of OceanX

I didn't love the ocean." In high school, Stewart became fascinated by public and global health, with a focus on environmental justice, but was unaware that marine science could be a viable career path. A chance encounter through social media led her to read about the Young Explorers Program and subsequently join an expedition. For ten days, she and a team of eleven other students from a diverse range of backgrounds traveled from the Azores to Norway onboard *OceanXplorer* to discover more about marine research and science communication. "What made this opportunity so cool was that there were scientists onboard doing actual work to publish in a paper. It wasn't staged like in an eleventh-grade science class," she says. "I got to help them out really closely, which was amazing and something I'd never experienced before. The crew didn't baby us; they treated us like we were one of them. We were always learning something and always having so much fun."

The Young Explorers were shown every element of the boat and how each department was run, and attended lectures each morning from scientists and mentors. Although Stewart had no prior experience and was nervous about coming aboard, the team brought her right into the research. On this particular mission, Matt Harke and Che Gardiner of the Gloucester Marine Genomics Institute planned to use *OceanXplorer*'s CTD instrument cluster to research a hydrothermal vent off the coast of the Azores. "When the vents ooze out hot water, it's like a plume, and they wanted to see how much biodiversity and nutrients came from that plume and how it supported the ecosystem," says Stewart. "I kind of became their right-hand man, and they were so supportive and affirming."

Stewart helped the team deploy the CTD to a depth of more than 2,000 meters (6,562 feet) and subsequently helped teach the other young explorers how to collect water samples from the CTD. "Up until that point in my life, I had never been that involved in science. I was actively doing things with my hands, and it was the coolest thing ever for me." While some of the Young Explorers had a

pre-existing passion for science, others were more inspired by storytelling, and learned to film, edit, and produce content for social media, all of which documented the research that the others were contributing to. "I was so impressed with the technology setup. The lighting, the cameras, everything had been so thought out," recalls Stewart. "It was just so evident that OceanX had put just as much time and money into the storytelling and technology aspect as they had into the science."

Programs like Young Explorers also highlight a generational shift in caring for the planet. Stewart notes that this shift can feel like a burden, but it is one she and others actively want to fight for. "I think my generation was born and raised with this immense pressure to fix the Earth. But I think a lot of us have a genuine passion for it. I think we're realizing that our current system of economy and society and culture is not sustainable for the planet or for human beings," she says. "A lot of us are going into climate and environmental science because we want to figure out ways to help the planet in a sustainable way." By offering opportunities to the next generation, OceanX Education is helping that change become a reality. Stewart is now studying marine biology. She wants to pursue postgraduate studies and a PhD in marine mammals and the ecology of coral and kelp ecosystems, all as a direct result of her journey onboard *OceanXplorer*. "I think marine science is just addictive. Marine scientists are not given the spotlight when talking about issues like climate change. I just think it's so important for people to know who they are, because just a little bit of knowledge can literally change your entire life. I became so confident after this experience, and I feel like I know myself better and know what I can do," she says. "I felt like I was with my best friends doing some incredible science in the most beautiful parts of the world. And there was such a healthy balance between work and fun. I would do it all again if I could."

OceanX also facilitates a host of educational experiences to engage with youth remotely. Classroom livestreams utilize the latest remote learning tools to share programming with students around the world, providing them with a one-on-one opportunity to learn about ocean

Left: A crustacean is studied up close in the wet labs onboard *OceanXplorer*.

> "Fifty years ago, Jacques Cousteau and *Calypso* managed to capture our imagination. Somehow, we lost that. We have to get back that personal connection to the sea."
>
> —**JAMES CAMERON**, filmmaker and explorer

exploration and conservation directly from experts in the field. Students hear from a diverse range of professionals about their research and experiences in a series of focused deep-dive conversations. The power of experiential learning creates a bridge between the vastness of the ocean and the curiosity of the human mind. Through innovative exhibits, OceanX makes marine science accessible, engaging, and inspiring for all, from students to educators to the general public.

With forecasts predicting that up to 90 percent of the world's coral reefs could disappear by 2050, the Reef Frame Your World campaign brought the wonders of coral reef systems to millions of New Yorkers through public transport displays. The amplification of this reached more than 25 million people. *Unseen Oceans*, a partnership with the American Museum of Natural History, brought the oceans to the public through an exhibition that utilized VR, robotics, and satellite monitoring to reveal the unseen habitats and creatures of the deep. In 2021, OceanX and Crayola collaborated to develop an ocean-themed exhibit in Orlando's Florida Mall. The exhibit included several child-friendly activities, including the creation and modeling of marine creatures on a 3D mapping table. A home version includes instructional activities and coloring experiences based on the adventures of the OceanX team.

The *OceanXperience* exhibition, a global traveling exhibition inspired by the research expeditions featured in the National Geographic series *OceanXplorers*, brings people onboard *OceanXplorer* through an interactive replica spread throughout a 300,000-square-foot gallery. Guests have the opportunity to simulate research and storytelling in a hyperrealistic format. One mission challenges them to study orca group dynamics by virtually piloting a drone to observe an orca pod, tagging an animal, and collecting audio and visual data to learn how members of the pod coexist. Other missions include journeys to research complex humpback whale communication and the tropical habitat of hammerhead sharks. As guests walk through the exhibition, they explore each section of *OceanXplorer* and have the opportunity to get a taste of what it's like to work on a research vessel. As Paul Hoffman, president and CEO of Liberty Science Center, noted, "The goal of the exhibition is not only to educate visitors of all ages, but also to connect them with the oceans in a way that inspires them to deeply engage with, understand, appreciate, and protect this majestic realm."

The education efforts have also been furthered by interactivity. Inspired by *Blue Planet II*, E-Line Media, in partnership with OceanX and BBC Studios, released the first-ever marine science video game, *Beyond Blue*. In

Right: The OceanX Education Portal and *Unseen Oceans* exhibit at the American Museum of Natural History provide immersive experiences with ocean content and new ways for future ocean advocates to learn.

HELO

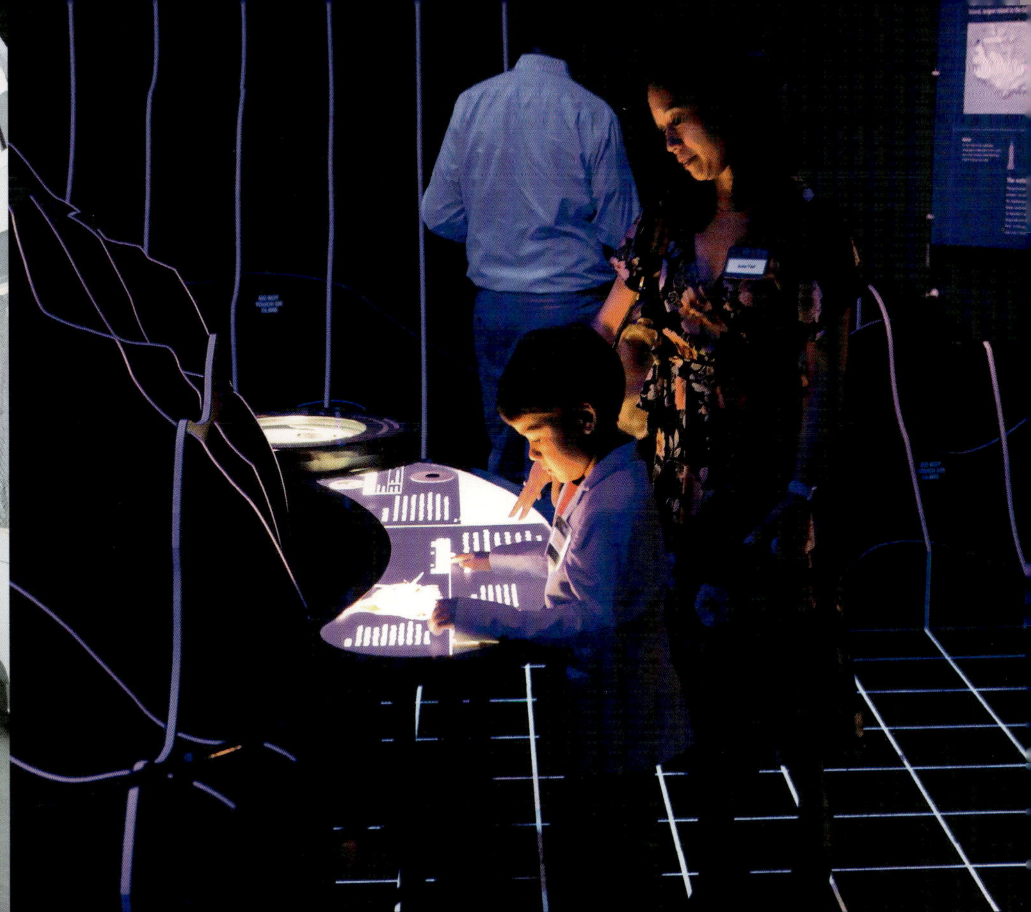

e animals!

TRITON

"My true passion lies in deep ocean science and in exploration. In *OceanXplorers*, we take viewers on an unparalleled ocean science adventure, introducing them to a group of principled, passionate people on a sometimes dangerous mission of ocean discovery in real time. We present the stakes, and firsts, and tell a visually spectacular and dramatic story—a story that inspires the next generation of explorers and adventurers committed to protecting and preserving our oceans."

—JAMES CAMERON, filmmaker and explorer

Beyond Blue, players don the wet suit of a marine scientist named Dr. Mirai Soto. The game mimics the experience of being on a research vessel like *OceanXplorer*, allowing players to navigate and study the habitats and animals of the ocean. Dr. Soto's ship acts as a home base, where samples and specimens from her adventures can be monitored and studied. By educating through play, *Beyond Blue* completely immerses gamers in a marine environment by highlighting some of the most dazzling features of the ocean.

From the moment they envisioned the *OceanXplorer*, Mark and Ray Dalio were imagining an epic documentary that only the new ship could facilitate. With its unique blend of scientific technology and media tools, the ship was a perfect platform to produce riveting ocean media. They connected with Director James Cameron, who shared their belief that showing the action, daring exploration, compelling science, and human drama of OceanX's missions held the potential to introduce a new audience to the wonder and potential of ocean exploration. The six-part National Geographic series *OceanXplorers* that resulted is the documentary the *OceanXplorer* vessel was built to create. Taking audiences on an immersive journey into the underwater world, the series filmed aboard the vessel over more than six months in 2021.

The series follows a charismatic and skilled team of researchers and innovators as they strive to learn more about ocean inhabitants and the pressures they face in a changing ocean environment. Esteemed scientists, including National Geographic Explorer and deep-sea researcher Zoleka Filander and shark biologist Melissa Cristina Márquez, and the experienced *OceanXplorer* crew, use the integrated exploration tools on the vessel to gather new insights from the ocean. Along the way, they uncover a unique amplification approach that humpback whales in the Dominican Republic use to attract potential mates. An incredible whale-tagging mission provides a first-person look at how deep-diving sperm whales hunt in the Azores. They uncover new details about hammerhead shark migration to the Bahamas and how this phenomenal animal navigates vast distances to reach this island paradise. And they travel north to the Svalbard archipelago to document how polar bears are adapting to their changing frozen world. And even farther north to the pack ice surrounding the North Pole, they encounter the world's longest-living vertebrae, the Greenland shark, in its natural habitat from a sub, becoming the first to observe its feeding strategies up close.

OceanXplorers premiered on National Geographic channels, Hulu, and Disney+ in September 2024. This unique series was a labor of love for the OceanX team, uniting ocean science, documentary filmmaking, and educational experiences in one global project. The series lives on, available on streaming services and through the OceanXperience traveling exhibition, which lets visitors recreate the studies featured in the series. Together, they show how incredible discoveries can push the boundaries of exploration and our understanding of the ocean to new frontiers.

Opposite (clockwise from top left): Ray Dalio and James Cameron chat aboard *OceanXplorer* during series filming.

OceanX crew secure a camera to *OceanXplorer's* helicopter for filming.

A producer descends to film in the media submersible, *Nadir*.

OCEANX

perisher

OCEAN

A REAL-TIME LOOK INSIDE OCEANX'S EXPLORATION

Covering every expedition, discovery, and education experience, OceanX's internal social media team has built an active, engaged community across Facebook, Instagram, LinkedIn, TikTok, X, and YouTube. This large global audience is the first to hear about OceanX's adventures, since the internal team embeds aboard the ship to cover the action in real time. Unlike traditional documentary filmmaking, which requires months of meticulous production, this integrated social media approach enables researchers and storytellers to share a behind-the-scenes look at OceanX's work, fostering authentic dialogue with global audiences, whether at sea or on land.

Authenticity is paramount for social media, with distinct storytelling approaches for each platform and an opportunity for followers to engage directly with OceanX. User comments and social media trends inform the OceanX content approach, helping what the organization publishes truly connect with their audience. Beyond the on-ship action, OceanX's social media feeds also tell the story of what the ocean needs to thrive and why a healthy ocean matters to humanity. Making this connection between OceanX's work and ocean sustainability not only reinforces the great need for this work, it also fosters a meaningful connection with its young, socially minded audience, who prioritize making a positive environmental impact. This transformation from the passive viewing of documentaries to dynamic participation of social media accelerates scientific understanding and policy momentum while democratizing marine exploration through direct public engagement.

Notable social media achievements include:

- 2.2B+ total impressions and 554M+ video views
- 10M+ followers across social platforms
- 75.5M audience engagements through 12,000+ social media posts
- Buzzworthy social campaigns like #squidtember reaching 21.8M people

Right: *OceanXplorer* helicopter approaches the ship in Raja Ampat, Indonesia.

Pages 184–185: An up-close image of a squid in *OceanXplorer*'s wet lab.

A WORLD OF DISCOVERY

As a research vessel, *OceanXplorer* has facilitated dozens of world-first missions and discoveries. During the maiden voyage to the Red Sea in 2021, a new species of black coral was discovered in the Neom region. Recently named *Bathypathes thermophilia*, the remarkable coral can survive in water temperatures of up to 22° Celsius and may shed light on the way organisms have adapted to thrive in warming waters. In the same region, the team spent upwards of 400 hours surveying the unexplored deep waters, which resulted in the discovery of four species of rays and nine species of sharks not documented in the Red Sea before.

BAHAMAS

2019: First to tag a bluntnose sixgill shark from a submersible at more than 500 meters deep.

Gavin Naylor for OceanX

SOUTH SHETLAND TRENCH, ANTARCTICA

2018: Completed the deepest human dive ever in Antarctica, to 1,000 meters, in a submersible.

OceanX

GULF OF AQABA, RED SEA

2020: Discovered deep-sea brine pools.
OceanX

JAPAN'S OGASAWARA ISLANDS

2012: Captured the first-ever footage of a giant squid in its natural habitat at a depth of 930 meters.
As seen on NHK/Discovery Channel / OceanX

RED SEA

2020: Discovered a new species of black coral.
OceanX

SEYCHELLES

2024: Conducted the deepest-ever ROV dive in the Amirantes Trench, reaching a depth of 4,600 meters.
OceanX

SOLOMON ISLANDS

2014: First to discover biofluorescence in more than 180 species of fish.
Courtesy of David Gruber

Further discoveries include the documentation of the shell-less mollusk *Firoloida desmarestia.* Though it has been observed in the epipelagic zone of many oceans, this was the first time it had been documented in the Red Sea. In the same year, while studying the biodiversity of hydrothermal vents in the waters surrounding the Azores, OceanX determined the adaptability of microbial populations—and gained a better understanding of how these unique ecosystems thrive—by studying their dispersal pattern near the vents.

In the summer of 2022, off the coast of Malta, an ancient amphora was unexpectedly discovered by university students participating in the Young Explorers Program while piloting the ROV at a depth of approximately 3,500 meters (11,483 feet). OceanX confirmed the discovery for the Republic of Malta and worked with the Malta Heritage Trust to safely preserve and deliver the rescued amphora to the University of Malta. The amphora was delivered in September 2023. In the United Arab Emirates in 2023, OceanX was the first to observe both mesophotic corals and the deep-sea bramble shark, along with collecting 783 biological samples of microorganisms from the water column. This research provided insights into the nutrients, environmental DNA, and temperatures of the region, which led to further understanding of the ecosystem and informed local fishery methods and conservation decisions. During the first-ever aerial survey of the Fujairah coastline in the western Gulf of Oman, the OceanX team made twelve sightings of megafauna, including bottlenose dolphins, spinner dolphins, striped dolphins, sea turtles, hammerhead sharks, and manta rays. The team also collected 112 samples of hard corals, soft corals, and invertebrates, many of which were entirely new to science.

Right: OceanX visited Malta in 2023 to return an ancient amphora discovered on the ocean floor.

Discoveries made in the Seychelles in 2024 represented a giant leap in ocean knowledge for the small nation. Utilizing the ROV and submersibles, the team partnered with local scientists to document deep-sea megafauna for the first time, along with mapping 12,000 square kilometers (4,633 square miles) of the Seychelle's exclusive economic zone. These discoveries not only add to the growing knowledge of the oceans, but also help the government of the Seychelles to make informed fishing and conservation decisions. Forever promoting science with state-of-the-art media and working to instill the curiosity and magic of ocean exploration in the public, OceanX live-streamed one of the submersible dives in the Seychelles to the World Economic Forum in Davos, where business leaders, ministers, and heads of state watched discoveries occur in real time.

Above: Controls aboard *OceanXplorer*.

Right: The crane aboard *OceanXplorer* supports submersible launches and retrievals.

This spread (clockwise from top right): An OceanX submersible captures biodiversity in Raja Ampat, Indonesia; during a Young Explorers program aboard *OceanXplorer*, a student discovered an ancient amphora on the ocean floor; an OceanX submersible captures biodiversity in Raja Ampat, Indonesia.

Pages 194–195: Mission control (left) and the wet lab (right) aboard *OceanXplorer*.

NO VIDEO
BlueView

OCEANX
OCEANX

1246

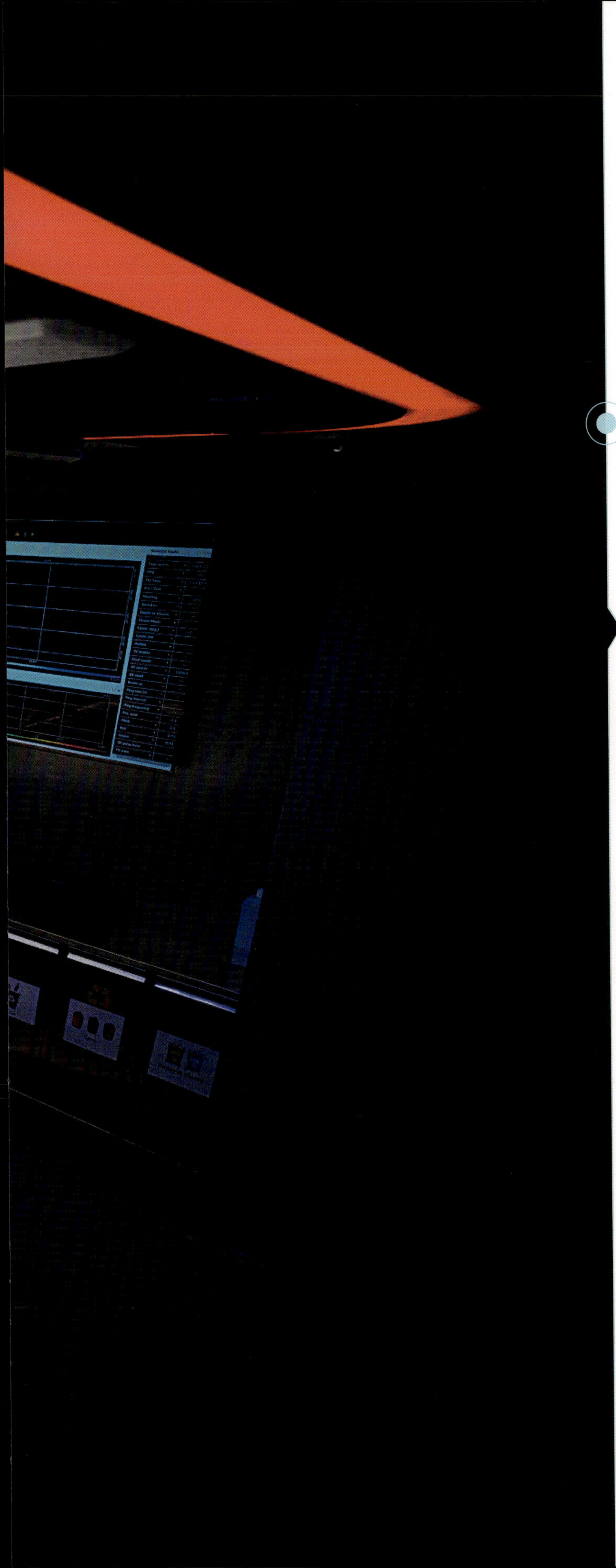

THE FUTURE OF *OCEANXPLORER*

Through collaboration with governments, research institutions, philanthropies, and nonprofit organizations, OceanX aims to utilize research and storytelling to generate ocean-conscious policy making. In this way, the future of OceanX is simple in its concept. Its objective is to work slowly and have *OceanXplorer* in one region for multiple years before exploring other ocean systems. This maximizes both the utilization of the ship and the ability to conduct comprehensive science. The big picture is not just to connect with individuals on a global scale, but to use the science and media to instill lasting change in the regions they visit. "The vessel will travel to all the large water bodies in the world one by one," says Vincent Pieribone. "If we're going to a place, we should help the country manage its resources. We ask them what they want to know and what would help them to manage their resources." Marine protected areas and strict extraction laws are important, but OceanX also understands that countries and economies need to thrive, and it intends to show countries sustainable ways in which to achieve both prosperity and conservation. As Pieribone states, "It's about people, it's about storytelling, it's about explaining that there are two paths, explaining that the future of the ocean is in your hands, and we think that future is valuable to you. It's not to say no to everything; let's do fishing and mining, but let's do it in a thoughtful manner. That's the whole sustainability concept."

Left: *OceanXplorer* submersible pilots review ship blueprint in mission control.

Pages 198–199: *OceanXplorer* off the coast of Svalbard in the Arctic Ocean. Photo by Taj Howe for OceanX.

PALFINGER
VALLETTA
CTD

OCEANXPLORER
ST-253 XR

In this vein, OceanX has been vocal on the international policy stage. The ocean is the world's largest carbon sink, however it is the least invested-in area of the UN's 2030 Agenda for Sustainable Development. As such, OceanX established itself at COP28, the 2023 UN Climate Change Conference held in Dubai, to provide data and facilitate discussions on marine research to develop comprehensive ocean management policies and educational initiatives. OceanX celebrated partnerships with ocean-centered organizations such as 1000 Ocean Startups (1000OS), a coalition of nearly fifty venture capitalists, business incubators, and other entrepreneur-supporting organizations, and an initiative of the World Economic Forum. The partnership with 1000OS provided ocean innovators the opportunity to test their technologies during research missions onboard *OceanXplorer*. OceanX's participation in COP28 coincided with its research mission in the UAE with partners from the Environment Agency—Abu Dhabi, M42, and Bayanat, and scientists from the Mubadala ACCESS Center at NYU Abu Dhabi. The UAE mission developed a comprehensive baseline under-standing of the region's ocean environment, along with conducting fishery assessments to create environmentally sustainable fishing tools and methods. As part of this, Dalio Philanthropies awarded the Mubadala ACCESS Center a $1 million grant to support coral research in the region.

Above: Mission control is the central nervous system of *OceanXplorer*, controlling, monitoring, and documenting all that happens below the service.

Right: Scientists observe mission findings aboard *OceanXplorer* during Mission Indonesia 2024.

In January 2024, OceanX partnered with the Environment Department of the Republic of Seychelles Ministry of Agriculture, Climate Change, and Environment, the Seychelles Islands Foundation, the Island Conservation Society, the University of Seychelles, the Seychelles Fishing Authority, and the Seychelles Conservation and Climate Adaptation Trust on a research mission to bridge the knowledge and data gap of the region's waters. They studied multiple aspects of the ecosystem, including deep-sea fish communities, seafloor habitat and structure, plankton communities, and megafauna. Since the Western Indian Ocean remains broadly unexplored, this expedition aimed to provide the government with valuable data to help protect the Seychelles' oceanic environment.

2024 also marked the start of a project in the Indo-Pacific region, including Malaysia, Indonesia, Singapore, the Philippines, and Thailand. "Southeast Asia is well known as one of the world's largest marine biodiversity hotspots, and much of it remains undiscovered and unexplored," says Mark Dalio. "There are few regions with a deeper connection to and dependence on the ocean. OceanX is committed to joining Southeast Asia in the battle against climate change by exploring the depths of its waters and bringing back critical data so we better understand how to protect and preserve it."

Left: Penguins leaping out of the ocean in Antarctica.

"The biggest problem facing the ocean is the very problem that OceanX is addressing. It's called ignorance. People don't know. You can't care if you don't know. We are all in this together: the squid, the cod, the plankton, the kelp, the coral reefs. We have to make peace with nature. That's why OceanX is so important."

—DR. SYLVIA EARLE, marine biologist and explorer

In addition, a partnership with Indonesia's Coordinating Ministry for Maritime and Investment Affairs, and the National Research and Innovation Agency, conducted scientific research and educational programming and developed inspiring content that promoted under-standing and care for the ocean. Research encompassed a wide range of crucial areas, including megathrust zone investigations to refine earthquake and tsunami models, fisheries research in Western Sumatra to en-hance understanding and guide management decisions, and exploration of potential biodiversity along with assessments of human impact. "Indonesia's ocean plays a central role in the country's economy, geopolitics, culture, and natural environment. With that in mind, OceanX is committed to providing valuable insights into this vital resource," says Pieribone. "By collecting comprehensive data, we aim to empower informed decision-making—for sustainable fishery management, effective conservation strategies, and proactive mitigation efforts to safeguard against future natural disasters." The mission also involved other Indonesian ministries and agencies, as well as researchers from universities and organizations such as Konservasi Indonesia.

Pages 204–205: Thriving reef ecosystems at Aldabra Atoll, Seychelles.

Right: Large coral reef photographed by scuba divers in Raja Ampat, Indonesia.

OCEANX PARTNERS

Philanthropic

- Tanoto Foundation
- TPC
- Rolex

Strategic

- Bloomberg Philanthropies
- Global Fund for Coral Reefs
- Pegasus
- The Prince Albert II of Monaco Foundation
- The Tiffany & Co. Foundation
- UNESCO's Intergovernmental Oceanographic Commission
- 1000 Ocean Startups

Scientific

- BRIN (National Research and Innovation Agency of Indonesia)
- CORDAP (Coral Research & Development Accelerator Platform)
- Gloucester Marine Genomics Institute
- Konservasi Indonesia
- National Geographic
- ORCA (Ocean Research and Conservation Association, Inc.)
- Scripps Institution of Oceanography
- National Oceanic and Atmospheric Administration
- Woods Hole Oceanographic Institution

In the years to come, *OceanXplorer* will continue to explore understudied and remote regions to not only gain a deeper understanding of the ocean, but also to share this with the world through captivating media, education programs, and policy dialogue, driving great progress for both the ocean and the people around the world who rely on it. The ocean's future—and ours—depends on this.

Right: OceanX team on mission in Norway.

Pages 210–211: A scuba diver closely films a majestic manta ray during a mission.

AFTERWORD: OUR VISION FOR A THRIVING OCEAN AND HUMANITY

Mark Dalio, OceanX Founder and co-CEO
Vincent Pieribone, OceanX co-CEO and Chief Science Officer

At OceanX, our commitment to the ocean is central to the organization. Since its inception, we have been driven by a deep belief that pushing the boundaries of ocean technology can unlock the knowledge to help create a vibrant future where the ocean and humanity thrive together in harmony. The *OceanXplorer* is proof of the power of technology to unlock this inspiring potential.

With the most advanced, multifaceted collection of ocean research and media capabilities on the planet, this vessel supports OceanX as we work to build a world where nature and humanity prosper together. Our scientific research and educational programs harness ocean knowledge to connect people to the ocean's value and potential. We believe these programs will help build the future we need.

Scientific research aboard the *OceanXplorer* is comprehensive, enabling a much better understanding of the ocean environment in record time. This understanding drives sustainable decision-making that balances ocean health with human needs. Ocean discoveries have—and will continue to—fuel economic pathways that support the fishing industry, global nutrition, technological innovations, and medical breakthroughs. Our collaborative approach, working with local scientists on local priorities and creating pathways to scientific collaboration, exponentially increases progress.

The ship also serves as an inspiring education platform, bringing the vision of a sustainable ocean one step closer every minute. The innovative media tools aboard the *OceanXplorer* amplify our science capabilities and fuel our education programs, bringing to life the ocean's wonders with creative, immersive content that connects a broader, more diverse community to the ocean. Whether through an onboard experience or OceanX Education's digital content, students and decision-makers understand why the ocean is worth protecting and experience the excitement of becoming part of the next generation of ocean advocates.

This exceptional vessel helps all who encounter it envision what is possible when humanity commits to ocean understanding. The *OceanXplorer* is the perfect platform for advanced scientific research, innovative educational opportunities, and innovative partnerships that will drive us toward an abundant ocean future. A future where the ocean is healthy and providing for humanity—plentiful food, growing economic potential, and astonishing solutions to human challenges. This ocean future is brighter than we can imagine. And it's possible when we push the boundaries of innovation, understanding, and connection. We hope you'll join us on our critical mission.

OCEANX
TRITON

PALFINGER
PALFINGER
OCEANXPLORER
MAJURO
OCEANXPLORER
MAJURO

Page 213: Ohe OceanX crew inspects a submersible before launch.

Left: The cranes aboard the *OceanXplorer* prepare to launch a submersible in the Azores.

Above: *OceanXplorer* docked for touring.

EARTH AWARE
An Imprint of Insight Editions
PO Box 3088
San Rafael, CA 94912
www.MandalaEarth.com

Publisher Raoul Goff
Associate Publisher Roger Shaw
Senior Editor Peter Adrian Behravesh
Assistant Editor Amanda Nelson
Managing Editor Michelle Hope
VP Creative Chrissy Kwasnik
Designer Amazing15
VP Manufacturing Alix Nicholaeff
Senior Production Manager Joshua Smith
Strategic Production Planner Lina s Palma-Temena

Earth Aware would also like to thank Maria Wilhelm, Ray Dalio, Mark Dalio, Vincent Pieribone, Alejandro Braceras, Andrew Craig, Carlos Fonseca, David Pollock, Eben Hall, Madeleine Stewart, Mario Tadinac, Mattie Rodrigue, Ollie Dolan, Roger Solem, Peter Fielding, Sam Purkis, Carly Tarricone, Rachel Rhodes, Connor Boals, Cassidy Nelson, Jermaine Harrell, Bob Cooper, Ian Shive, and Kelly Henthorne.

ISBN: 978-1-64722-541-4

Manufactured in China by Insight Editions
10 9 8 7 6 5 4 3 2 1

ROOTS of PEACE
REPLANTED PAPER

Insight Editions, in association with Roots of Peace, will plant two trees for each tree used in the manufacturing of this book. Roots of Peace is an internationally renowned humanitarian organization dedicated to eradicating land mines worldwide and converting war-torn lands into productive farms and wildlife habitats. Roots of Peace will plant two million fruit and nut trees in Afghanistan and provide farmers there with the skills and support necessary for sustainable land use.